W0050723

Plasma Cell Neoplasms

Robert B. Lorsbach · Marwan Yared
Editors

Plasma Cell Neoplasms

Pathogenesis, Diagnosis and Laboratory
Evaluation

 Springer

Editors
Robert B. Lorsbach
Department of Pathology and Laboratory
  Medicine
Cincinnati Children's Hospital Medical
  Center, University of Cincinnati College
  of Medicine
Cincinnati, OH
USA

Marwan Yared
Department of Pathology and Immunology
Baylor College of Medicine
Houston, TX
USA

ISBN 978-3-319-42368-5   ISBN 978-3-319-42370-8   (eBook)
DOI 10.1007/978-3-319-42370-8

Library of Congress Control Number: 2016945872

© Springer International Publishing Switzerland 2016
This work is subject to copyright. All rights are reserved by the Publisher, whether the whole or part of the material is concerned, specifically the rights of translation, reprinting, reuse of illustrations, recitation, broadcasting, reproduction on microfilms or in any other physical way, and transmission or information storage and retrieval, electronic adaptation, computer software, or by similar or dissimilar methodology now known or hereafter developed.
The use of general descriptive names, registered names, trademarks, service marks, etc. in this publication does not imply, even in the absence of a specific statement, that such names are exempt from the relevant protective laws and regulations and therefore free for general use.
The publisher, the authors and the editors are safe to assume that the advice and information in this book are believed to be true and accurate at the date of publication. Neither the publisher nor the authors or the editors give a warranty, express or implied, with respect to the material contained herein or for any errors or omissions that may have been made.

Printed on acid-free paper

This Springer imprint is published by Springer Nature
The registered company is Springer International Publishing AG Switzerland

Contents

Contributors

Cerisse Harcourt Myeloma Institute for Research and Therapy, University of Arkansas for Medical Sciences, Little Rock, AR, USA

Pei Lin Department of Hematopathology, University of Texas-MD Anderson Cancer Center, Houston, TX, USA

Robert B. Lorsbach Department of Pathology and Laboratory Medicine, Cincinnati Children's Hospital Medical Center, University of Cincinnati College of Medicine, Cincinnati, OH, USA

L. Nicholas Cossey Nephropath, Little Rock, AR, USA

Soumya Pandey Department of Pathology, University of Arkansas for Medical Sciences, Little Rock, AR, USA

Jeffrey R. Sawyer Department of Pathology and Myeloma Institute, University of Arkansas for Medical Sciences, Little Rock, AR, USA; Cytogenetics Laboratory, Little Rock, AR, USA

Shree G. Sharma Nephropath, Little Rock, AR, USA

Giampaolo Talamo Penn State Hershey Cancer Institute, Hershey, PA, USA

Marwan Yared Department of Pathology and Immunology, Baylor College of Medicine, Houston, TX, USA

Maurizio Zangari Myeloma Institute for Research and Therapy, University of Arkansas for Medical Sciences, Little Rock, AR, USA

Chapter 1
Introduction

Marwan Yared and Robert B. Lorsbach

Overview

The plasma cell neoplasms are clonal proliferations of the most terminally differ-entiated stage of B-cell differentiation, the plasma cell (PC), and include a family of neoplasms, the prototype of which is plasma cell myeloma (PCM). These are common hematolymphoid neoplasms, with over 25,000 new cases of PCM diag-nosed annually in the United States [1]. As with their benign counterpart, a defining feature of PC neoplasms is their synthesis of immunoglobulin, in the form of a monoclonal immunoglobulin molecule referred to as an M-protein or paraprotein. In addition to end-organ damage mediated directly by the neoplastic PCs them-selves, the deposition of M-protein, especially that occurring in the kidney, accounts for much of the morbidity and mortality associated with these neoplasms.

Epidemiologic and biologic investigations have now confirmed that virtually all cases of PCM arise from an antecedent, clinically occult state termed monoclonal gammopathy of undetermined significance or MGUS [2, 3]. MGUS is most com-monly detected in individuals greater than 50 years of age, with an incidence of approximately 3% [4]. Although incompletely understood, the incidence of MGUS is influenced by ethnicity and other genetic factors, resulting in a significantly higher incidence in African-Americans and lower incidence in Asians compared to individuals of European ancestry [5]. As with PCM, the incidence of MGUS

M. Yared (✉)
Department of Pathology and Immunology, Baylor College of Medicine, 1502 Taub Loop—702, Houston, TX 77030, USA
e-mail: marwan.yared@bcm.edu

R.B. Lorsbach
Department of Pathology and Laboratory Medicine, Cincinnati Children's Hospital Medical Center, University of Cincinnati College of Medicine, 3333 Burnet Ave., ML 1035, Cincinnati, OH 45229, USA
e-mail: robert.lorsbach@cchmc.org

© Springer International Publishing Switzerland 2016
R.B. Lorsbach and M. Yared (eds.), *Plasma Cell Neoplasms*,
DOI 10.1007/978-3-319-42370-8_1

increases with advancing age. Although virtually all PCM arises in the setting of MGUS, the rate at which MGUS progresses to PCM is low, approximately 1% per year.

By contrast, a significantly higher fraction of patients with asymptomatic (smoldering) PCM progresses to symptomatic disease necessitating therapeutic intervention. In one study of asymptomatic PCM patients, the median time to progression was 4.8 years, with a risk of progression of 10% per year in the first five years after diagnosis, 3% in the next five years, and 1% in subsequent years [6]. A major question in the myeloma field is whether the prospective identification of those patients with asymptomatic PCM (or even MGUS) who are at very high risk for disease progression is feasible and, if so, whether therapeutic intervention in such patients at an earlier stage of disease and at a younger age might reduce morbidity and improve clinical outcomes, possibly with less intense therapy [7].

As will be evident from our discussion below, the diagnosis and final classification of the PC neoplasms nearly always requires the integration of clinical, radiologic, pathologic, and laboratory findings. The detection and quantitation of serum paraprotein levels play an important role in both the clinical diagnosis of PC neoplasms as well as in assessing their response to therapy; the kinetics of serum paraprotein reduction in response to therapy are prognostically important [8, 9]. For the pathologist, assessment of immunoglobulin expression by immunohistochemistry and/or flow cytometry is critical for confirming the clonality of a bone marrow PC population and has diagnostic utility for the distinction of PCM from other B-cell lymphomas such as lymphoplasmacytic lymphoma.

Classification of plasma cell neoplasms

Plasma cell neoplasms as defined in the 2008 World Health Organization Classification of Tumours of Haematopoietic and Lymphoid Tissues consist of plasma cell myeloma (multiple myeloma), plasmacytoma, monoclonal immunoglobulin deposition diseases (primary amyloidosis and light and heavy chain deposition diseases), and osteosclerotic myeloma (POEMS syndrome) [10]. MGUS is considered a precursor lesion and is therefore included in this discussion.

Monoclonal Gammopathy of Undetermined Significance

MGUS is defined as the presence of a serum monoclonal protein of <30 g/L in patients who have <10% clonal plasma cells in the bone marrow and who entirely lack any evidence of myeloma-related end-organ damage, often referred to by the mnemonic CRAB (hypercalcemia, renal insufficiency, anemia, and lytic bone lesions). There should also be no evidence of involvement by any other disease that might be associated with a paraproteinemia, in particular B-cell lymphoma. MGUS is more prevalent in men than women (1.5:1 male:female ratio) and is detected in about 3% of persons older than 50 years and 5% of persons older than 70 years [10]. The rate of progression to multiple myeloma is reported to be about 1% per year [11–13].

Plasma Cell Myeloma

PCM is defined as a PC neoplasm that is associated with production of an M-protein in serum and/or urine. PCM comprises 1% of all malignancies and 15% of hematological neoplasms. It is more prevalent in males (1.4:1 male:female ratio) and the median age at diagnosis is 70 years. The disease is almost always bone marrow-based and is often widely disseminated in the bone marrow at initial diagnosis. Extramedullary disease may be present at initial presentation and is relatively common in the latter stages of the disease.

According to the 2008 WHO classification of hematopoietic and lymphoid neoplasms, the following three criteria must be met to render a diagnosis of symptomatic PCM:

1. Presence of an M-protein in serum or urine (no minimal protein level required for the diagnosis)
2. Presence of neoplastic clonal PCs in the bone marrow (no minimal percentage required for the diagnosis), or plasmacytoma
3. Myeloma-related organ or tissue impairment as evidenced by any of the following:

   a. CRAB (hypercalcemia, renal insufficiency, anemia, or lytic bone lesions)
   b. Hyperviscosity
   c. Anemia
   d. Recurrent infections (see Table 1.1)

Table 1.1 2008 WHO Diagnostic criteria for MGUS, smoldering (asymptomatic) myeloma, and plasma cell myeloma (symptomatic)

2008 WHO	MGUS	Smoldering myeloma (asymptomatic)	Plasma cell myeloma (symptomatic)
Bone marrow clonal plasma cells (% of nucleated cells)	<10%	≥10%	Any percentage (but usually ≥10%) or plasmacytoma
Serum M-protein (g/L)	<30	Myeloma levels	Any level (but usually ≥30 of IgG or ≥25 of IgA or 1 g/24 h of urine light chain)
Myeloma-related organ or tissue damage or symptoms	None	None	CRAB or hyperviscosity or amyloidosis or recurrent infections

Three clinical variants of PCM are recognized by the WHO, namely, asymptomatic (smoldering) PCM, non-secretory PCM, and plasma cell leukemia. A diagnosis of asymptomatic (smoldering) PCM requires the detection of a serum or urine M-protein and confirmation of a clonal, neoplastic PC population in the bone marrow or a plasmacytoma in an otherwise asymptomatic patient with no evidence of myeloma-related end-organ damage. Only a small subset of PCM patients, approximately 8%, are diagnosed with asymptomatic PCM, 73% of whom progress to symptomatic PCM or primary amyloidosis within 15 years of initial diagnosis [6]. Non-secretory PCM is defined by the absence of any detectable M-protein as evaluated by immunofixation electrophoresis. As described in Chapter 3, many cases of "non-secretory" PCM retain some low-level paraprotein biosynthetic capacity when evaluated by newer, more sensitive techniques (e.g., serum free light chain analysis) and are thus best considered hyposecretory rather than true non-secretory PCM. Of note, non-secretory PCM has a lower incidence of extramedullary manifestations like renal insufficiency, hypercalcemia, and hypogammaglobulinemia. Finally, the last PCM variant is plasma cell leukemia (PCL), a subtype defined by the presence of circulating clonal PCs in the peripheral blood exceeding $2 \times 10^9/L$ or comprising $\geq$20% of leukocytes [4]. Primary PCL is an uncommon variant, representing only about 2–5% of all cases of PCM. Secondary PCL occurs most often in the agonal phase of the disease in the setting of a heavy disease burden. Extramedullary manifestations like lymphadenopathy, organomegaly, and renal failure are more prevalent in PCL [14].

In November of 2014, the International Myeloma Working Group (IMWG) published updated criteria for the diagnosis of symptomatic and asymptomatic PCM [15]. According to the updated IMWG, the following two criteria must be met to render a diagnosis of symptomatic PCM:

1. Clonal bone marrow plasma cells $\geq$10% or plasmacytoma
2. Presence of one of the following myeloma-defining events

    a. CRAB (hypercalcemia, renal insufficiency, anemia, or lytic bone lesions)
    b. Clonal bone marrow plasma cell percentage $\geq$60%
    c. Involved: uninvolved serum free light chain ratio of $\geq$100
    d. Two or more focal lesions on MRI studies

According to the updated IMWG, the following two criteria must be met to render a diagnosis of smoldering (asymptomatic) PCM:

1. Serum monoclonal protein (IgG or IgA) $\geq$30 g/L or urinary monoclonal protein $\geq$500 mg per 24 h, and/or clonal bone marrow plasma cells 10–60%
2. Absence of myeloma-defining events or amyloidosis (see Table 1.2)

Table 1.2 2014 IMWG Diagnostic criteria for MGUS, smoldering (asymptomatic) myeloma, and plasma cell myeloma (symptomatic)

2014 IMWG	MGUS	Smoldering myeloma (asymptomatic)	Plasma cell myeloma (symptomatic)
Bone marrow clonal plasma cells (% of nucleated cells)	<10%	10–60%[a]	≥10% or Plasmacytoma
Serum M-protein (g/L)	<30	≥30 or Urine M-protein ≥500 mg/24 h[a]	Not required for diagnosis
Myeloma-defining events	None	None	CRAB or BM clonal plasma cells ≥60% or Serum FLC ratio≥100 or ≥2 focal lesions on MRI

[a]In addition to absence of any myeloma-defining event, presence of clonal BM plasma cells (10–60%) and/or a serum or urine M-protein is required for a diagnosis of smoldering myeloma

Plasmacytoma

Plasmacytoma is a tumor comprised of monoclonal plasma cells with no evidence of myeloma-related organ or tissue impairment, other than localized osseous damage [10]. A monoclonal protein is detected in up to 72% of patients [16]. Plasmacytomas comprise 3–5% of plasma cell neoplasms and occur more frequently in males (male: female ratio of 2:1) at a median age of 55 years [17, 18]. Plasmacytoma can be localized to the bone (solitary plasmacytoma of bone) or be extramedullary (extraosseous plasmacytoma). The diagnosis of solitary plasmacytoma of bone requires a complete skeletal survey lacking any evidence of other osseous lesions. The persistence of a detectable M-protein in patients with osseous plasmacytoma following radiotherapy has strong negative impact, with the majority of such patients subsequently developing PCM [19]. An obvious diagnostic consideration with extraosseous plasmacytoma that must be excluded is B-cell lymphoma with extensive plasmacytic differentiation.

Monoclonal Immunoglobulin Deposition Diseases

Per the 2008 WHO classification of hematopoietic and lymphoid neoplasms, the monoclonal immunoglobulin deposition diseases (MIDD) include primary amyloidosis as well as the monoclonal light and heavy chain deposition diseases. Primary amyloidosis is a PC neoplasm that results in the deposition of abnormal immunoglobulin molecules in various tissues and end organs ultimately resulting in

Table 1.3 Characteristics of monoclonal immunoglobulin deposition diseases (primary amyloidosis and light and heavy chain deposition diseases), and heavy chain diseases

	Primary amyloidosis	Monoclonal light and heavy chain deposition diseases	Heavy chain diseases
Type of disorder	Plasma cell neoplasm	Plasma cell neoplasm	B-cell neoplasm
Disorders	Primary Amyloidosis	LCDD HCDD LHCDD	Alpha HCD Gamma HCD Mu HCD
Protein deposition	Yes	Yes	No
Amyloid formation	Yes	No	No

tissue damage and organ dysfunction [20]. The immunoglobulin deposits may consist of intact immunoglobulin molecules or of isolated light chain or rarely heavy chain molecules. The abnormal immunoglobulin molecules assume a β-pleated sheet configuration and avidly bind the diazo dye, Congo red, which shows apple-green birefringence when examined under polarized light. Most cases of primary amyloidosis occur in the setting of MGUS; however, approximately 20% of patients have concomitant PCM. Primary amyloidosis occurs more frequently in males (male: female ratio of 2:1) at a median age of 64 years [21, 22].

Similar to primary amyloidosis, monoclonal light and heavy chain deposition diseases are PC neoplasms. In light chain deposition disease (LCDD) an abnormal immunoglobulin light chain is secreted and deposited in various tissues causing tissue damage. Less often, an abnormal heavy chain or both a light and a heavy chain are secreted and deposited in heavy chain deposition disease (HCDD) and light and heavy chain deposition disease (LHCDD), respectively. Unlike primary amyloidosis, the abnormal protein secreted in these diseases does not have a β-pleated sheet configuration, is not congophilic, and thus lacks the apple-green birefringence characteristic of amyloid. Monoclonal light and heavy chain deposition diseases occur with equal frequency in males and females at a median age of 56 years. Of note, monoclonal light and heavy chain deposition diseases must be distinguished from heavy chain diseases which are B-cell neoplasms that cause production, but not deposition of abnormal heavy chains (see Table 1.3).

References

1. Siegel RL, Miller KD, Jemal A. Cancer statistics, 2015. CA Cancer J Clin. 2015;65:5–29.
2. Landgren O, Kyle RA, Pfeiffer RM, et al. Monoclonal gammopathy of undetermined significance (MGUS) consistently precedes multiple myeloma: a prospective study. Blood. 2009;113:5412–7.
3. Weiss BM, Abadie J, Verma P, Howard RS, Kuehl WM. A monoclonal gammopathy precedes multiple myeloma in most patients. Blood. 2009;113:5418–22.

4. Kyle RA, Buadi F, Rajkumar SV. Management of monoclonal gammopathy of undetermined significance (MGUS) and smoldering multiple myeloma (SMM). Oncology (Williston Park, NY) 2011;25:578–86.

5. Landgren O, Weiss BM. Patterns of monoclonal gammopathy of undetermined significance and multiple myeloma in various ethnic/racial groups: support for genetic factors in pathogenesis. Leukemia. 2009;23:1691–7.

6. Kyle RA, Remstein ED, Therneau TM, et al. Clinical course and prognosis of smoldering (asymptomatic) multiple myeloma. New Engl J Med. 2007;356:2582–90.

7. Caers J, Fernandez de Larrea C, Leleu X, et al. The changing landscape of smoldering multiple myeloma: a European perspective. Oncologist. 2016;21:333–42.

8. Schaar CG, Kluin-Nelemans JC, le Cessie S, Franck PF, te Marvelde MC, Wijermans PW. Early response to therapy and survival in multiple myeloma. Br J Haematol. 2004;125:162–6.

9. Shah J, Blade J, Sonneveld P, et al. Rapid early monoclonal protein reduction after therapy with bortezomib or bortezomib and pegylated liposomal doxorubicin in relapsed/refractory myeloma is associated with a longer time to progression. Cancer. 2011;117:3758–62.

10. Swerdlow SH, Campo E, Harris NL, et al. WHO classification of tumours of haematopoietic and lymphoid tissues. Lyon: IARC Press; 2008.

11. Kyle RA, Therneau TM, Rajkumar SV, et al. A long-term study of prognosis in monoclonal gammopathy of undetermined significance. N Engl J Med. 2002;346:564–9.

12. Cesana C, Klersy C, Barbarano L, et al. Prognostic factors for malignant transformation in monoclonal gammopathy of undetermined significance and smoldering multiple myeloma. J Clin Oncol: Off J Am Soc Clin Oncol. 2002;20:1625–34.

13. Montoto S, Blade J, Montserrat E. Monoclonal gammopathy of undetermined significance. New Engl J Med. 2002;346:2087–8; author reply-8.

14. Garcia-Sanz R, Orfao A, Gonzalez M, et al. Primary plasma cell leukemia: clinical, immunophenotypic, DNA ploidy, and cytogenetic characteristics. Blood. 1999;93:1032–7.

15. Rajkumar SV, Dimopoulos MA, Palumbo A, et al. International myeloma working group updated criteria for the diagnosis of multiple myeloma. Lancet Oncol. 2014;15:e538–48.

16. Criteria for the classification of monoclonal gammopathies. multiple myeloma and related disorders: a report of the International Myeloma Working Group. Br J Haematol. 2003;121:749–57.

17. Alexiou C, Kau RJ, Dietzfelbinger H, et al. Extramedullary plasmacytoma: tumor occurrence and therapeutic concepts. Cancer. 1999;85:2305–14.

18. Dimopoulos MA, Moulopoulos LA, Maniatis A, Alexanian R. Solitary plasmacytoma of bone and asymptomatic multiple myeloma. Blood. 2000;96:2037–44.

19. Wilder RB, IIa CS, Cox JD, Weber D, Delasalle K, Alexanian R. Persistence of myeloma protein for more than one year after radiotherapy is an adverse prognostic factor in solitary plasmacytoma of bone. Cancer. 2002;94:1532–7.

20. Merlini G, Seldin DC, Gertz MA. Amyloidosis: pathogenesis and new therapeutic options. J Clin Oncol: Off J Am Soc Clin Oncol. 2011;29:1924–33.

21. Kyle RA, Gertz MA. Primary systemic amyloidosis: clinical and laboratory features in 474 cases. Semin Hematol. 1995;32:45–59.

22. Gertz MA. Immunoglobulin light chain amyloidosis: 2014 update on diagnosis, prognosis, and treatment. Am J Hematol. 2014;89:1132–40.

Chapter 2
Clinical Features, Management, and Therapy of Plasma Cell Neoplasms: What Pathologists Need to Know

Giampaolo Talamo, Cerisse Harcourt and Maurizio Zangari

Introduction

Plasma cell neoplasms are characterized by the proliferation of monoclonal plasma cells and include a spectrum of disorders with different clinical manifestations, treatment, and prognosis: monoclonal gammopathy of undetermined significance (MGUS), smoldering myeloma (SM), symptomatic multiple myeloma (MM), solitary plasmacytoma (SP) of the bone, extramedullary plasmacytoma (EMPC), plasma cell leukemia (PCL), monoclonal immunoglobulin deposition diseases (MIDD) including primary (AL) amyloidosis and light and heavy chain deposition diseases, and osteosclerotic myeloma/POEMS syndrome. The basic features of these clinical entities are shown in Table 2.1.

In this chapter, we will focus primarily on multiple myeloma (MM), which is the most common malignant plasma cell disorder. MM accounts for 1% of all cancers, and is the second most common hematologic malignancy, with approximately 22,000 new cases and 11,000 deaths per year in the United States. [1]. The age-adjusted incidence is 6:100,000 per year in the United States and Europe. Median age at diagnosis is 69 years, and three-quarters of patients are older than 55 years; only 2% of patients are younger than 40 years [2]. Multiple myeloma is slightly more predominant in men than women (3:2), and more common in African Americans, with an incidence two to three times higher than in the Caucasian population [3].

G. Talamo
Penn State Hershey Cancer Institute, Hershey, PA, USA
e-mail: gtalamo@hmc.psu.edu

C. Harcourt · M. Zangari (✉)
Myeloma Institute for Research and Therapy, University of Arkansas for Medical Sciences, 4301 West Markham, Slot 816, Little Rock, AR, USA
e-mail: mzangari@uams.edu

© Springer International Publishing Switzerland 2016
R.B. Lorsbach and M. Yared (eds.), *Plasma Cell Neoplasms*,
DOI 10.1007/978-3-319-42370-8_2

Table 2.1 Main characteristics of plasma cell disorders

	Paraprotein	Clonal plasma cells in bone marrow	Clinical manifestations	Treatment
MGUS	+	<10%	No	Observation
Smoldering myeloma	+	≥10%	No	Observation (or clinical trial)
Multiple myeloma	+	≥10%	CRAB	Chemotherapy ASCT Palliative RT
Solitary plasmacytoma of the bone	±	<10%	Bone lesion	RT (potentially curable)
Extramedullary plasmacytoma	±	No	Depending on affected organ	RT (potentially curable)
Plasma cell leukemia	+	>10%, with >20% or >2000/mL circulating PCs	CRAB	Chemotherapy ASCT
AL amyloidosis	FLC	Usually detected by flow cytometry	Depending on affected organ	Chemotherapy ASCT
Monoclonal immunoglobulin deposition disease	Usually kappa	±	Depending on affected organ	Chemotherapy ASCT
POEMS	Usually lambda	<10%	Polyneuropathy, Castleman disease, sclerotic bone lesions	Chemotherapy ASCT

Abbreviations BM: bone marrow; CRAB: hypercalcemia, renal insufficiency, anemia, bone lesions; FLC: free light chain; MGUS: monoclonal gammopathy of undetermined significance; RT: radiation therapy; ASCT: stem cell transplant

Clinical Features of Plasma Cell Neoplasms

Virtually all cases of MM are preceded by an MGUS phase [4, 5], a condition in which monoclonal proteins (IgG, IgA, kappa light chains, or lambda light chains) can be detected in the blood and/or urine, usually secreted by a small (<10%) population of clonal plasma cells in the bone marrow. MGUS is a relatively common medical condition, affecting approximately 3% of the population over 50 years of age [6]. A few patients with MGUS present with symptoms of peripheral neuropathy, usually with tingling and numbness in the hands and legs, an increased risk of venous and arterial thrombosis, infections, and osteoporosis. However, the vast majority have no specific symptoms and a normal life expectancy. The risk of progression to symptomatic MM from MGUS is approximately 1% per year, and 15% in a lifetime [7]. It is important to recognize that rarely patients with IgM MGUS evolve into IgM MM: the vast majority of these patients are at risk of progression to lymphoplasmacytic lymphoma/Waldenström macroglobulinemia [8]. MGUS does not require treatment. Patients are usually followed at 6-month intervals with laboratory testing including paraprotein levels, and if stable, followed annually.

Patients with a paraprotein and ≥10% clonal plasma cells in their bone marrow, but no clinical manifestations (CRAB criteria), are classified as smoldering mye- loma. The bone marrow aspirate/biopsy (≥10% clonal plasma cells) and the amount of the serum paraprotein (usually >3 g/dL) differentiate these patients from those with MGUS. In patients with smoldering myeloma, the risk of progression to symptomatic MM is higher than in MGUS: approximately 10% per year in the first 5 years, and 75% in a lifetime [9]. Patients with MGUS and smoldering myeloma are generally asymptomatic. They are usually incidentally diagnosed when a paraprotein is detected during the screening and workup of other medical problems, such as anemia, renal insufficiency, peripheral neuropathy, or osteoporosis. A bone marrow aspirate/biopsy is necessary for the final diagnosis. The percentage of clonal plasma cells in the bone marrow is usually <10 and ≥10% in MGUS and smoldering myeloma, respectively. Not every patient with MGUS needs a bone marrow aspirate/biopsy. Some experts feel it could be avoided if the M protein is of the IgG subtype if the serum concentration is ≤1.5 g/dL. The chance of detecting symptomatic MM in these patients is negligible [10]. It is important to note that the definition of smoldering myeloma has recently changed. Patients with asymp- tomatic MM who have >60% plasma cells in the bone marrow (2–3% of cases), or those with a serum involved/uninvolved free light chain ratio ≥100 (15% of cases), are now defined as symptomatic MM and considered for therapy. Retrospective clinical studies have shown that almost all of these patients inevitably develop signs and symptoms of MM within several months of the diagnosis [11, 12]. According to the current International Myeloma Working Group (IMWG) guidelines, two criteria must be met to establish the diagnosis of smoldering myeloma: (1) serum M protein (IgG or IgA) ≥3 g/dL or urine M protein >500 mg/24 h, and/or 10–60% clonal plasma cells in the bone marrow and (2) absence of CRAB symptoms (see below) or amyloidosis [13].

Pathologists should be aware that the absence of bone marrow plasmacytosis does not exclude the diagnosis of MM, because occasionally the bone marrow may not be involved both at diagnosis and in relapsed cases [14, 15]. In our experience, among 586 patients with an established diagnosis of MM who were followed in 2010–2014 at the Penn State Hershey Cancer Institute, the bone marrow aspirate and biopsy at baseline contained <10% clonal plasma cells in 53 (9%) of these patients and the bone marrow was negative for plasma cells, even by flow cytometry, in 14 of them (2%) (unpublished data).

The main clinical manifestations of MM are summarized by the CRAB symp- toms: hypercalcemia, renal insufficiency, anemia, and bone lesions. A significant proportion of MM patients can present with several other clinical manifestations due to a variety of pathogenetic mechanisms (Table 2.2): (a) clone related such as lytic lesions, hypercalcemia, and pancytopenia due to diffuse infiltration of the bone marrow; (b) mechanical complications due to tumor mass pressure causing spinal cord and nerve root compression; (c) extramedullary involvement with organ specific symptoms; (d) malignant plasma cells can secrete paraproteins and other

Table 2.2 Clinical manifestations of multiple myeloma at presentation [90]

Diffuse infiltration of the bone marrow	
• Anemia	58%
• Leukopenia	N/A
• Thrombocytopenia	N/A
Bone destruction	
• Hypercalcemia	9%
• Painful osteolytic lesions	53%
• Pathologic fractures	9%
Mass effect	
• Spinal cord compression	3%
• Nerve root compression	2%
Plasma cells outside of bone marrow	
• Plasma cell leukemia	2%
• Extramedullary disease	4%
Secretion of paraproteins/other molecules	
• Cast nephropathy/renal insufficiency	17%
• Hyperviscosity syndrome	3%
• Amyloidosis	2%
• Cryoglobulinemia	1%
• DKK1-mediated osteoporosis	N/A
Other	
• Peripheral neuropathy	7%
• Coagulopathy	2%
• Infections	1%
• Systemic symptoms	2%
• Secondary gout	1%

molecules causing plasma hyperviscosity, cast nephropathy, amyloid deposition, cryoglobulinemia, and DKK1-mediated osteoporosis.

Multiple Myeloma has the greatest incidence of *bone involvement* among all hematological malignancies [16]. Macroscopic bone lesions are present in 80% of MM patients and are typically osteolytic. Osteolytic lesions can lead to pathologic fractures (i.e., spontaneous as opposed to posttraumatic) and bone deformities (Figs. 2.1 and 2.2). Bone lesions seldom heal in MM, even in patients who achieve complete remission, and, the persistence of lytic lesions on conventional radiographs cannot be used as a tool to monitor response to therapy. Newer and more sophisticated imaging techniques, such as Magnetic Resonance Imaging (MRI) and the positron emission tomography with 2-deoxy-2-[18] fluoro-D-glucose (FDG-PET) (Fig. 2.3) indicate areas with metabolically active lesions and, therefore, are more useful.

Renal insufficiency in MM is usually associated with cast nephropathy, where the kappa or lambda light chains bind to the urinary Tamm–Horsfall protein (THP) and form obstructing casts in the distal tubules [17]. THP is a low-molecular-weight glycoprotein synthesized by the thick ascending limb of the Henle's loop. Each kappa

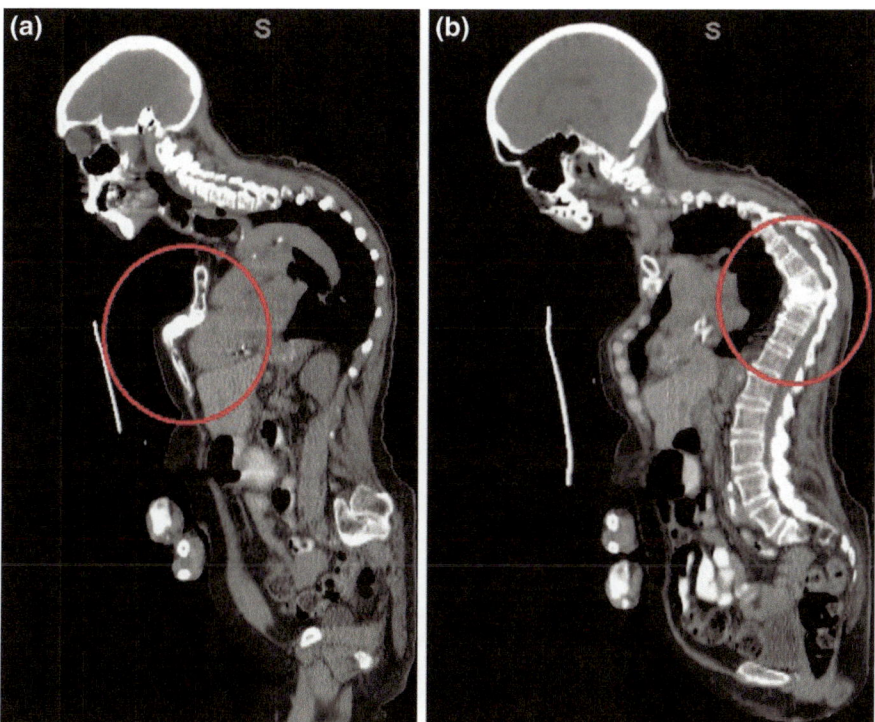

Fig. 2.1 Multiple myeloma can produce painful bone fractures. These CT scans show severe deformity of the chest wall, due to a sternal fracture (**a**), and a compression fracture of the spine causing kyphosis (**b**)

or lambda light chain binds to the THP through the complementary determining region (CDR) [18], the affinity of which is affected by the amino acid sequence. Cast nephropathy formation is related to the presence of elevated kappa or lambda free light chain (usually >100 mg/dL) with a significant affinity for the THP. Although cast nephropathy is the most common cause of renal failure in MM, other conditions should be considered in these patients, including hypercalcemia, hyperuricemia, direct plasma cell infiltration of the renal parenchyma, amyloid deposition, crystal storing histiocytosis, and pyelonephritis. Acquired Fanconi syndrome is another renal manifestation of MM due to tubular dysfunction induced by the filtered light chains and is characterized by hypouricemia, hypokalemia, hypophosphatemia, glycosuria, normal anion gap, metabolic acidosis, amino aciduria, and osteomalacia [19]. The most useful clues for diagnosing this entity are the presence of hypouricemia and glycosuria in a patient with normal serum glucose.

Hyperviscosity syndrome occurs when the elevated paraproteins cause an increase in blood viscosity: typically with IgG > 7000 mg/dL and IgA > 5000 mg/dL. The most common clinical manifestations are: (a) mucosal bleeding, including epistaxis, spontaneous ecchymoses, and gum bleeding; (b) blurred vision and visual loss. An

Fig. 2.2 MRI imaging summarizing the complications of MM involving the spine: compression fractures of several vertebral bodies, severe kyphotic deformity, and spinal cord compression

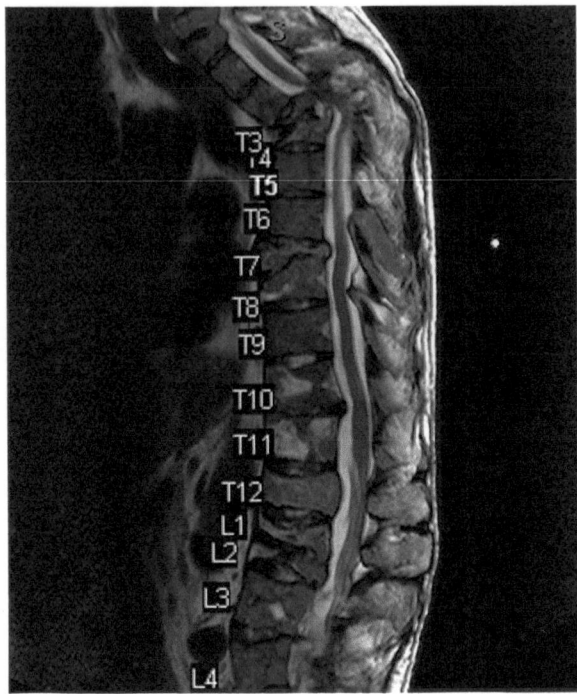

opthalmoscopic exam may detect dilated retinal veins, with sausage-like segmentation, and retinal hemorrhages; (c) neurologic symptoms including headache, vertigo, ataxia, paresthesias, somnolence, confusion, and others; and (d) cardiorespiratory symptoms including dyspnea, hypoxia, congestive heart failure, volume overload, lower extremity edema, pulmonary edema, and elevated jugular venous pressure.

Extramedullary plasmacytomas (EMPCs) are plasma cell neoplasms that arise outside the bone marrow. The most frequent location of primary EMPCs is the upper respiratory tract, especially the nasal fossa and the paranasal sinuses. EMPCs (Extramedullary plasmacytomas) in the context of MM develop in about 5% of patients at the time of the diagnosis, and eventually develop in 20% of patients during the course of the disease. EMPCs (Extramedullary plasmacytomas) may involve different tissues, including pleura, pericardium, gastrointestinal mucosa system, genitourinary tract, skin (Fig. 2.4), and central nervous system, usually in the form of meningeal myelomatosis. While solitary EMPCs are potentially curable neoplasms, the development of extramedullary involvement in patients with MM is associated with a highly aggressive clinical course and poor outcomes [20, 21].

Plasma Cell Leukemia (PCL) is defined by an absolute plasma cell count of at least 2000/mL in the peripheral blood, or >20% plasma cells of the total white cell count. If present at the time of MM diagnosis, it is called *primary* PCL. *Secondary* PCL is the leukemic transformation of a previously diagnosed MM, and it usually develops in the terminal phase of the disease. Clinical manifestations of PCL are

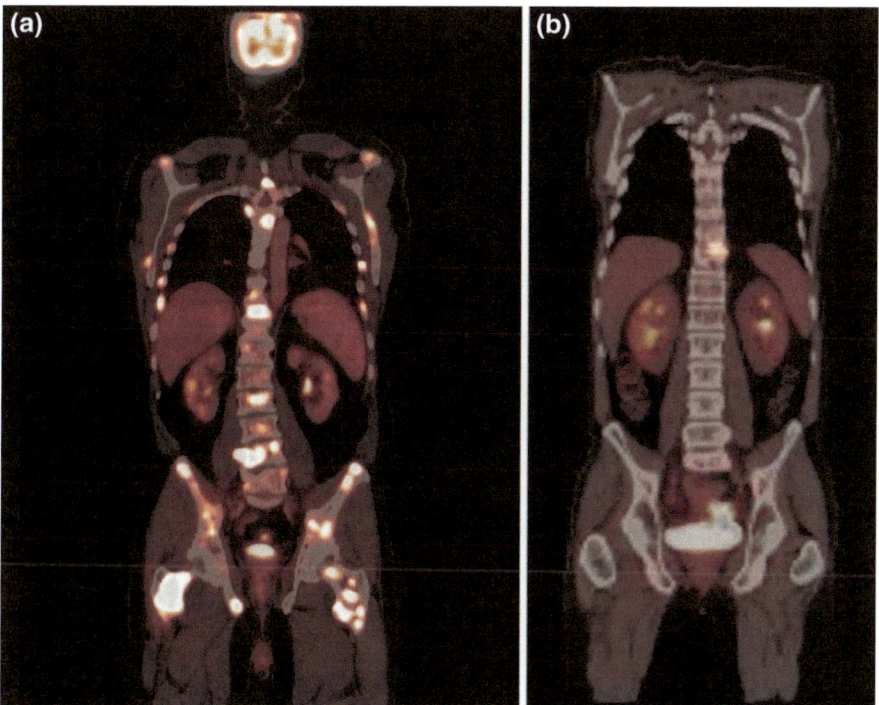

Fig. 2.3 Positron emission tomography with 2-deoxy-2-[18]fluoro-D-glucose (FDG-PET), indicating the metabolically active bone lesions (**a**). In a newly diagnosed patient with only a single lesion (**b**), the bone marrow biopsy will determine whether the plasma cell disorder is multiple myeloma (usually incurable) or a solitary plasmacytoma of the bone (potentially curable)

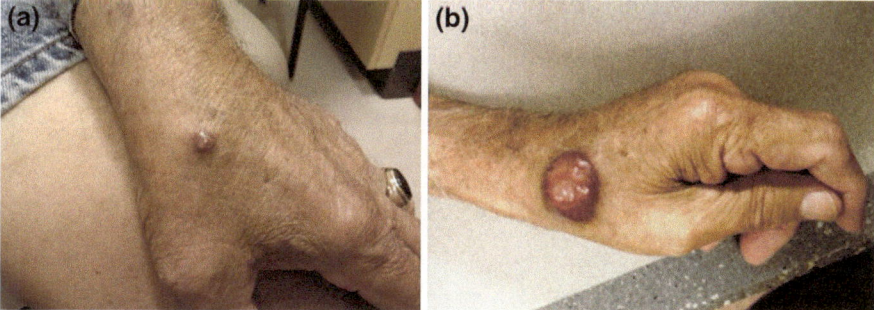

Fig. 2.4 Extramedullary involvement of the skin in a patient with multiple myeloma. The lesion enlarged over several weeks

similar to those of MM, but these patients may also have splenomegaly and hepatomegaly. The clinical outcome of PCL is significantly worse than in MM, and survival is often measured in weeks or months instead of years [22–24].

In *amyloidosis*, the kappa or lambda light chains produced by the malignant plasma cell clone precipitate in the organs and form the characteristic β-pleated fibrils, which appear birefringent on Congo red staining when viewed under a polarized microscope. Amyloidosis can occur as part of MM manifestations, but it may also develop in the absence of MM features. In this case, it is called primary AL amyloidosis, and it should be distinguished from other nonmalignant forms of the disease, such as secondary amyloidosis (for example, chronic inflammatory conditions, such as rheumatoid arthritis), or the hereditary form (transthyretin—TTR—amyloidosis). AL amyloidosis is produced by a hypoproliferative clone of plasma cells in the bone marrow that shed amyloidogenic light chains of either kappa, or more frequently, lambda type. On clinical grounds, the presence of amyloidosis should be suspected when there are certain manifestations, such as nephrotic-range albuminuria, autonomic or peripheral neuropathy including carpal tunnel syndrome, hepatomegaly, coagulopathy (factor X consumption), skin purpura due to fragile blood vessels (Fig. 2.5 a), macroglossia (Fig. 2.5 b), diarrhea, and malabsorption due to the involvement of the GI tract, xerostomia due to infiltration of the minor salivary glands, congestive heart failure, and sudden cardiac death due to arrhythmias. A typical finding in patients with cardiac amyloidosis is the presence of low voltages on the ECG despite left ventricular hypertrophy seen at the echocardiogram (Fig. 2.5).

Monoclonal Light and Heavy Chain Deposition Diseases are diseases with a clinical phenotype similar to that of AL amyloidosis, and are associated with MM in approximately 65% of cases [25]. The immunoglobulin molecules can form deposits in various organs as in amyloidosis, but they do not form fibrils, and they do not show the apple-green birefringence on Congo red staining. The immunoglobulin light chains can deposit either alone or in combination with the heavy chains, or deposits can be composed of heavy chains only. As in amyloidosis, the organ most commonly involved by monoclonal light where heavy chain deposition diseases is the kidney, and patients usually present with varying degrees

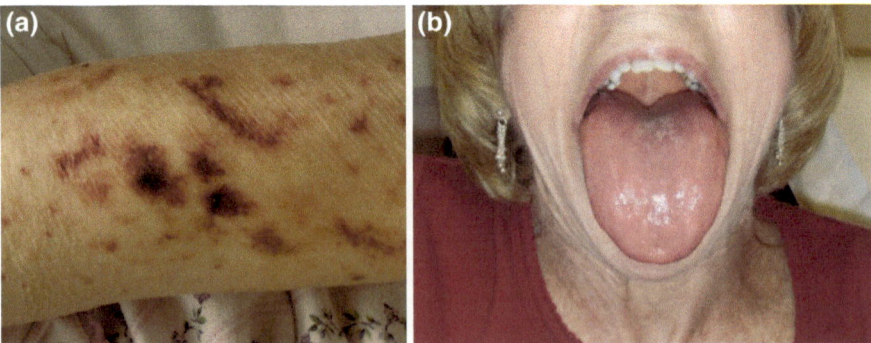

Fig. 2.5 Purpuric/ecchimotic lesions of the skin (**a**) and macroglossia (**b**) in patients with primary (AL) amyloidosis

of proteinuria and renal insufficiency. Other organs such as the heart, the liver, and the lung may also be involved.

POEMS is a mnemonic for *p*olyneuropathy, *o*rganomegaly, *e*ndocrinopathy, *M* protein, and *s*kin changes. However, in this syndrome, not all features of the mnemonic need to be present, and patients with this syndrome have other manifestations not included in the mnemonic, such as sclerotic bone lesions, extravascular fluid overload, papilledema, erythrocytosis, thrombocytosis, and elevated vascular endothelial growth factor (VEGF) levels. The diagnosis of POEMS syndrome is usually made on the basis of monoclonal plasma cells in an osteosclerotic lesion. Updated diagnostic criteria have been published and include two mandatory criteria (polyneuropathy and the presence of a monoclonal plasma cells, almost always lambda-restricted), and at least one of the three major features (Castleman disease, sclerotic bone lesions, and VEGF elevation) [26].

Diagnostic Evaluation

Detection of Paraproteins and Laboratory Tests

Each immunoglobulin molecule has two heavy chains (IgG, IgA, IgD, IgE, and IgM) and two light chains (either kappa or lambda). The monoclonal immunoglobulin secreted by the malignant plasma cells, called "M protein", can be used as a tumor marker, not only to diagnose the presence of the disease, but also to follow the disease course over time and the response to therapy [27]. The type of the heavy and light chains compromising the M protein can be identified by IFE (Immunofixation Electrophoresis), and can be quantitated by agarose gel protein electrophoresis (PEL) and capillary zone electrophoresis [28, 29]. IFE is not a quantitative test, but it is more sensitive than PEL, and therefore is more useful for identifying minimal residual disease. In about 20% of cases, MM cells produce only light chains, either kappa or lambda. In these cases, the M protein cannot be detected by serum PEL (the light chains have a low molecular weight and are rapidly cleared into the urine), but circulating free lights can be detected and monitored with the free light chain (FLC) assay. Guidelines by the IMWG recommend the use of all three serum assays—IFE, PEL, and FLC—for the diagnosis of plasma cell disorders, because of the assays' different specificities and sensitivities [30]. Historically, the standard test to quantify the M protein in the urine was the 24 h urine collection. However, not only is this method inconvenient and inaccurate [31], but it does not increase the sensitivity when the FLC assay is included in the diagnostic panel [32]. Current IMWG guidelines do not recommend urine tests for diagnosis, unless amyloidosis is suspected (because renal involvement in this disease can be manifested by significant albuminuria) [30]. However, the PEL of the 24 h urine is still required for monitoring the response to therapy in patients with measurable urine M proteins [33]. Of note, in about 1% of cases, MM is nonsecretory, and patients do not have M proteins in the serum or in the urine.

Other laboratory tests that are useful in the diagnosis and follow-up of patients with MM are the complete blood count (CBC), which may reveal cytopenias, the complete metabolic profile (CMP), which may reveal hypercalcemia and renal insufficiency, albumin, and beta-2-microglobulin (which reflects both tumor mass and renal insufficiency), for assessing the ISS stage. An elevated level of lactate dehydrogenase (LDH) correlates with aggressive disease and short survival [34], and it has been integrated in the recently published Revised International Staging System (R-ISS). Since MM is one of the secondary causes of osteoporosis, a vitamin D level may be useful.

Bone Marrow Aspirate and Biopsy

A bone marrow aspirate and biopsy is required in all patients with MM, because the percentage of plasmacytosis is essential for the diagnosis, and also because important prognostic information can be generated by metaphase cytogenetic analysis and fluorescence in situ hybridization (FISH) (see "Prognosis and clinical outcomes") performed on the bone marrow specimen. MM cells usually appear as atypical plasma cells: large-sized, binucleated, with hyperchromatic chromatin, and with prominent nucleoli. Bartl et al. have described a morphologic classification that includes six histologic types and three prognostic grades: low grade ("Marschalko" and "small cell" types), intermediate grade ("cleaved", "polymorphous", and "asynchronous" types), and high grade ("blastic" type) [35]. Phenotyping, by immunohistochemical (IHC) staining or flow cytometric analysis shows that MM cells are usually positive for CD38, CD138, and CD56, and negative for CD19 and CD20 [36]. Flow cytometric analysis of the bone marrow aspirate also has several important clinical applications, including the differential diagnosis of MM versus MGUS versus reactive plasmacytosis, the diagnosis of nonsecretory MM, the detection of minimal residual disease after therapy [37], and the detection of early relapse [38].

Imaging Studies

The traditional method to assess the number and extent of bone lesions in MM is a *skeletal survey*, which includes radiographs of the skull, chest, spine, pelvis, humeri, and femora. Signs of MM are punched-out lytic lesions, osteoporosis, and pathologic fractures. However, plain radiographs have several limitations: they are normal in 15–20% of patients, and they are relatively insensitive, because they do not detect osteolytic lesions unless >30% of trabecular bone is lost. Moreover, they are not useful to assess response to therapy and documenting disease remission, because lytic bone lesions rarely heal in MM, even when patients are in complete remission [39]. More sophisticated methods of imaging are CT scans, MRI, and PET/CT scans. *CT scans* are useful in detecting small osteolytic lesions invisible on plain X-rays, and allow better visualization of some bones (e.g., sternum, ribs, scapulae) compared to skeletal surveys. The main advantage of the *MRI* is its increased sensitivity in detecting bone disease: in fact, bone marrow infiltration may be detected on MRI in about 40% of patients with negative X-rays, and, therefore, MRI is indicated in all patients with normal skeletal surveys [40]. Magnetic Resonance Imaging also allows the visualization of soft tissue masses extending from the bone lesions, and is the technique of choice to detect spinal cord compression or nerve compression. Typically, the vertebral bodies are hyperintense (i.e., brighter than the intervertebral disks) on T1-weighted images, and hypointense (i.e., darker than the intervertebral disks) on fat-suppressed T2-weighted images. If the bone marrow cellular component increases and fatty component decreases, the signal decreases on T1 and increases on T2 and STIR (=short time inversion recovery) images. Myeloma lesions are hypointense on T1, hyperintense on T2, and hyperintense on STIR with enhancement on postcontrast T1 sequence. This pattern is useful for distinguishing MM lesions from vertebral body hemangiomas, which are hyperintense on T1 and T2 sequences due to fat content, and their high signal disappears on the STIR sequence. There are several patterns of marrow infiltration on MRI: "focal", "diffuse", and "salt and pepper". *Positron emission tomography* with 2-deoxy-2-[18]fluoro-D-glucose (FDG-PET) imaging is useful in MM for several reasons. It provides a whole body image, shows sites of extramedullary disease, and, more importantly, is able to differentiate between old inactive bone lesions and new active ones, which is particularly useful when evaluating treatment response (on X-rays and MRI, inactive lesions may persist and lead to a false-positive results). Studies have shown that persistent positivity of PET/CT after chemotherapy predicts worse prognosis by multivariate analysis [41, 42]. PET scans are of particular value in patients with nonsecretory MM and in patients with solitary plasmacytoma. Bone lesions should be at least 5–10 mm in size to be visualized by PET. Skeletal MRI and PET/CT scans are not interchangeable. In general, MRI is usually superior to a PET/CT in detecting MM lesions, whereas PET/CT is more sensitive in assessing response to therapy. Dual-energy X-ray absorptiometry (*DEXA*) scans should be considered in all patients with MM, because it can indicate the presence of osteopenia/osteoporosis, with associated risk of fractures, and also monitor response to therapeutic interventions [43, 44] (Table 2.3).

Table 2.3 Main adverse prognostic factors in multiple myeloma

Clinical features	• Renal failure (creatinine >2 mg/dL) • Extramedullary disease • Plasma cell leukemia
Laboratory	• Low albumin, high β2 m (ISS stage) • High LDH level
Metaphase cytogenetics	• Hypodiploidy • Monosomy 13
FISH	• 17p− • 1q21+, 1p− • t(4;14), t(14;16)

Treatment of Plasma Cell Neoplasms

Survival of MM patients has significantly improved in the last two decades. The benefit mainly occurred after the introduction of high-dose chemotherapy and autologous stem cell transplantation (ASCT), especially in younger patients, and the use of novel agents at relapse [45]. ASCT has become a standard of care as part of upfront therapy for younger patients. Achieving complete remission (CR) post-ASCT is the crucial step for long-lasting response. However, it is now clear that the quality of the response to pre-transplant induction influences survival as well. Higher response rates to pre-transplant induction have been obtained with three drug combinations, in particular the regimens with bortezomib, thalidomide, and dexamethasone (VTD), cyclophosphamide, bortezomib, and dexamethasone (VCD or CyBorD), and lenalidomide, bortezomib, and dexamethasone (RVD) [3]. Post-transplant, consolidation, and maintenance interventions have been developed to extend the duration of the response and ultimately, the overall survival (OS). Consolidation is usually administered 2–4 cycles after transplant, and maintenance includes a reduced-intensity treatment on a long-term basis in the attempt to stabilize the previously achieved response. Patients not eligible for ASCT with symptomatic disease and organ damage (i.e., hypercalcemia, real failure, anemia, or bone lesions) also benefit from immediate treatment. The Revised International Staging System (R-ISS) and chromosomal abnormalities define high- and standard-risk patients. Proteasome inhibitors, immunomodulatory drugs, corticosteroids, and alkylating agents are the most active agents. All these chemotherapy drugs have different mechanisms of action, and are directed against various cellular targets in the MM cells (Fig. 2.6). The presence of comorbidities and the performance status should be thoroughly assessed, and treatment should be tailored to individual patients. Bone disease, renal damage, hematologic toxicities, infections, thromboembolism, and peripheral neuropathy are the most frequent complications requiring prompt and specific supportive care measures.

When a plasma cell neoplasm is localized, as in the case of SP of the bone and EMPC, chemotherapy is not necessary, and these tumors respond well to involved field radiation therapy, with 40–70% of cases achieving definitive cure [46]. Surgery is required in only a minority of cases of EMPC [47].

Monoclonal gammopathy of undetermined significance and smoldering myeloma patients do not need therapy, mainly because they do not manifest end organ damage. These patients are generally asymptomatic and may remain stable for many years. A recent clinical trial involving the use of lenalidomide and dexamethasone in a subset of patients with smoldering myeloma and high-risk features has demonstrated a clinical benefit with prolonged time to progression and survival advantage [48]. However, this approach is not routinely adopted in clinical practice, and should be done only in the context of clinical trials, because the definition of high-risk smoldering myeloma in the aforementioned trial was based on a multiparametric flow cytometry method that has not been standardized.

Younger patients presenting with symptomatic MM are treated with systemic chemotherapy. Induction therapy in newly diagnosed patients can be followed by an ASCT in eligible patients. Timing of transplant is a matter of controversy. Some experts advocate for transplant later during the course of the disease ("salvage" ASCT, as opposed to ASCT at first remission). Some studies indicated an ASCT performed at relapse led to a similar OS compared to an early ASCT [49, 50]. Regardless of the timing of ASCT, patients with MM can currently have an excellent 4-year survival of >80% [50]. Although chemotherapy, with or without ASCT, can lead to long-term remission, sometimes lasting several years, definitive cure is seen in only a minority of patients, and relapses are the rule. Therefore, patients with MM typically undergo multiple lines of chemotherapy, in various sequence and combinations, often with two or more agents. A typical MM patient undergoes multiple lines of therapy, with phases of remissions and relapses. With each successive regimen, depth, and duration of response progressively diminishes, until the disease becomes refractory to all available therapies [51]. The assessment of the response to therapy has been standardized, and follows the updated criteria established by the IMWG [27]. These criteria, shown in Table 2.4, include progressive disease (PD), stable disease (SD), minor response (MR) partial response (PR), very good partial response (VGPR), complete response (CR), and stringent complete response (sCR).

MM and other plasma cell neoplasms are sensitive to a variety of cytotoxic drugs. A list of the most commonly used agents is shown in Table 2.5.

Corticosteroids are among the most commonly used drugs in the treatment of MM. They have been shown to induce response rates of approximately 60% in newly diagnosed patients when used as monotherapy [52].

Alkylating drugs, such as melphalan and cyclophosphamide, induce the formation of crosslinks between two strands of DNA ("interstrand cross-linking"), with impairment of DNA synthesis and cell replication. Melphalan, a drug synthesized in 1953, has a structure that incorporates two alkylating agents, nitrogen mustard, and phenylalanine. More than 60 years later, this drug is still considered standard therapy for MM, either as single agent in the preparative regimen for autologous ASCT, or in combination with other agents in ASCT-ineligible patients [53].

Immunomodulatory Drugs (IMiDs). Thalidomide and its derivatives lenalidomide and pomalidomide represent a class of oral antineoplastic compounds called ImmunoModulatory Drugs. Until recently, the receptor protein for these drugs was unknown, but a recent study demonstrated that IMiDs bind to the protein Cereblon (CRBN), a component of a ubiquitin ligase called cullin 4 ring ligase complex (CRL4) [54]. Therefore, IMiDs should be more properly classified as "ubiquitin ligase modulators". The therapeutic activity of the IMiD lenalidomide is due to a CRBN gain of function which modifies the substrate specificity of CRBN, enabling the IMiD-bound CRBN to induce the proteasome degradation of the Ikaros family zinc finger proteins IKZF1 and IKZF3. These are two B cell transcription factors, that can be transcriptional activators or repressors, depending on different cellular

Table 2.4 Revised uniform IMWG criteria to assess response to therapy in multiple myeloma [27]

Response category	Response criteria
Progressive disease (PD)	Increase of 25% from lowest response value in any of the following: • Serum M component with absolute increase ≥0.5 g/dL; serum M component increases ≥1 g/dL are sufficient to define relapse if starting M component is ≥5 g/dL and/or • Urine M component (absolute increase must be ≥200 mg/24 h) and/or • Only in patients without measurable serum and urine M protein levels: difference between iFLC and uFLC levels (absolute increase must be >10 mg/dL) • Only in patients without measurable serum and urine M protein levels and without measurable disease by FLC level, BM PC % (absolute % must be ≥10%) Development of new or definite increase in size of existing bone lesions or soft tissue plasmacytomas Development of hypercalcemia that can be attributed solely to plasma cell proliferative disorder Two consecutive assessments before new therapy are needed
Stable disease (SD)	Not meeting criteria for CR, VGPR, PR, or PD; no known evidence of progressive or new bone lesions if radiographic studies were performed
Minimal response (MR) (for relapsed refractory myeloma only)	≥25% but ≤49% reduction of serum M protein and reduction in 24 h urine M protein by 50–89% In addition, if present at baseline, 25–49% reduction in size of soft tissue plasmacytomas is also required No increase in size or number of lytic bone lesions (development of compression fracture does not exclude response)
Partial response (PR)	PR ≥50% reduction of serum M protein and reduction in 24-hour urinary M protein by ≥90% or to <200 mg/24 h If serum and urine M protein are not measurable, ≥50% decrease in difference between iFLC and uFLC levels is required in place of M protein criteria If serum and urine M protein and serum FLC assay are not measurable, ≥50% reduction in BM PC is required in place of M protein, provided baseline % was ≥30% In addition, if present at baseline, ≥50% reduction in size of soft tissue plasmacytomas is required Two consecutive assessments are needed; no known evidence of progressive or new bone lesions if radiographic studies were performed
Very good partial response (VGPR)	Serum and urine M component detectable by immunofixation but not on electrophoresis or ≥90% reduction in serum M component+urine M component <100 mg/24 h; in patients for whom only measurable disease is by serum FLC level, >90% decrease in difference between iFLC and uFLC levels, in addition to VGPR criteria, is required; two consecutive assessments are needed
Complete response (CR)	Negative immunofixation of serum and urine, disappearance of any soft tissue plasmacytomas, and <5% PC in BM; in patients for whom only measurable disease is by serum FLC level, normal FLC ratio of 0.26–1.65 in addition to CR criteria is required; two consecutive assessments are needed

(continued)

Table 2.4 (continued)

Response category	Response criteria
Stringent complete response (sCR)	CR as defined+normal FLC ratio and absence of clonal PCs by immunohistochemistry or 2- to 4-color flow cytometry; two consecutive assessments of laboratory parameters are needed
Immunophenotypic CR	sCR as defined+absence of phenotypically aberrant PCs (clonal) in bone marrow with minimum of 1 million total BM cells analyzed by multiparametric flow cytometry (with >4 colors)
Molecular CR	CR as defined+negative allele-specific oligonucleotide PCR (sensitivity 10^{-5})

Abbreviations BM: bone marrow; CR: complete response; FLC: free light chain; iFLC: involved free light chain; M: monoclonal; MR: minimal response; PC: plasma cell; PCR: polymerase chain reaction; PD: progressive disease; PR: partial response; sCR: stringent complete response; SD: stable disease; uFLC: uninvolved free light chain; VGPR: very good partial response

settings [55, 56]. Thalidomide, initially introduced in Germany in 1957 as a sedative, was withdrawn from the market in 1961, when it was linked to severe fetal malformations. The drug was recently reintroduced after the discovery of its anti-myeloma activity in patients with relapsed disease [57, 58]. Thalidomide analogues lenalidomide and pomalidomide are second-generation IMiDs, developed to enhance anticancer properties and reduce the adverse effects associated with thalidomide. Lenalidomide proved effective in refractory patients, including those who had relapsed following thalidomide treatment [59], and has become one of the most frequently prescribed drugs for MM. Pomalidomide elicited responses in 47% of patients who had received three or more previous regimens, including lenalidomide [60].

Proteasome Inhibitors. The proteasome is a protein enzyme complex that breaks down and clears unused or misfolded proteins. MM cells are particularly sensitive to inhibition of the proteasome, presumably because they are specialized for the mass production of immunoglobulins. The increased protein load associated with this task lowers the threshold for proteotoxic stress and leaves plasma cells susceptible to toxic misfolded/unfolded proteins and proapoptotic signals, initiated by the unfolded protein and endoplasmic reticulum (ER) stress responses [61]. The 26S proteasome consists of two 19S regulatory complexes and a barrel-shaped 20S proteolytic core. The 19S regulatory complexes bind the proteins tagged with ubiquitin and direct them to the 20S core. The 20S is a proteolytic core that consists of 2α-subunit rings and 2β-subunit rings, each of which contains seven different α and β subunits. The proteolysis is mediated by the β-subunits: β1 (caspase-like activity), β2 (trypsin-like activity), and β5 (chymotrypsin-like activity).

The proteasome inhibitor bortezomib is a dipeptide boronate approved by the U. S. Food and Drug Administration (FDA) in 2003, after its clinical efficacy was demonstrated in refractory MM [62]. After its remarkable success in the treatment of relapsed MM, it was later approved for first-line treatment. As single agent, bortezomib induces responses in approximately one-third of patients with relapsed MM, and the median duration of response is 1 year [62, 63].

Table 2.5 Drugs commonly used in the treatment of multiple myeloma

		Comments	Typical adverse reactions
Corticosteroids	• Dexamethasone • Prednisone	Commonly used in combination regimens with other agents	Insomnia, agitation hypertension, hyperglycemia, AVN, weight gain, myopathy, fluid retention
Conventional chemotherapy agents	• Melphalan	Used orally in elderly patients, and IV as the most common conditioning regimen for ASCT	Cytopenias, N/V, mucositis, diarrhea, alopecia
	•	Cyclophosphamide	Used in low dose in combination regimens, and high dose as mobilization therapy before ASCT
	Cytopenias, N/V, mucositis, diarrhea, alopecia		
	• Bendamustine	Usually given in the advanced phase of the disease	Cytopenias, N/V, diarrhea
• Doxorubicin	Most commonly used formulation is the pegylated liposomal doxorubicin	Cytopenias, N/V, mucositis, diarrhea, alopecia	
Proteasome inhibitors	• Bortezomib	SC administration has same efficacy but less neurotoxicity than IV	Peripheral neuropathy, N/V, diarrhea
	• Carfilzomib	Usually given after failure of bortezomib and an IMiD	Fatigue, N/V, cytopenias, diarrhea, fever
IMiDS	• Thalidomide	First-generation IMiD, less potent than lenalidomide and pomalidomide	Sedation, sensory peripheral neuropathy, constipation, VTE, teratogenicity
	• Lenalidomide	Most commonly used IMiD	Cytopenia, skin rash, VTE, teratogenicity, secondary malignancies
	• Pomalidomide	Usually given after failure of bortezomib and another IMiD	Cytopenias, peripheral neuropathy, VTE, teratogenicity
HDAC inhibitors	• Panobinostat	Given with bortezomib, after failure of bortezomib and an IMiD	Diarrhea, fatigue, nausea, cytopenias
Monoclonal antibodies	* Daratumumab	Usually given after failure of bortezomib and another IMiD	Infusion-related reactions
	* Elotuzumab	Usually given in combination with lenalidomide	Infusion-related reactions

Abbreviations AVN: avascular necrosis of the bone; HDAC: histone deacetylase; IMiD: immunomodulatory drug; IV: intravenous; N/V: nausea/vomiting; SC: subcutaneous; ASCT: stem cell transplant; VTE: venous thromboembolism

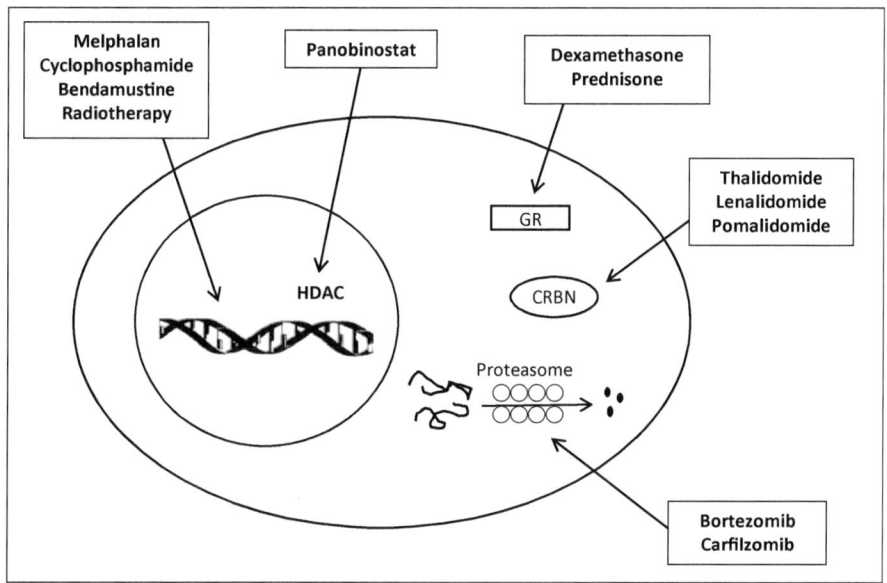

Fig. 2.6 Cellular targets of chemotherapeutics agents in myeloma cells. Melphalan, cyclophosphamide, and Bendamustine are alkylating agents disrupting the DNA. Radiation therapy primarily exerts its cytotoxic damage by altering the DNA structure. Panobinostat is a histone deacetylase inhibitor. Corticosteroid hormones bind to the cytoplasmic glucocorticoid receptor (GR), which migrates in the nucleus and alters DNA transcription. Thalidomide, lenalidomide, and pomalidomide bind to cerebron (CRBN), and modify the function of ubiquitin ligases. Bortezomib and carfilzomib inhibit the proteasome, the organelle which digest cellular proteins

Carfilzomib, a second-generation proteasome inhibitor, has also shown significant clinical efficacy against relapsed MM [64]. The activity of carfilzomib against bortezomib-resistant MM cells may be due to its pharmacological profile, which differs from bortezomib. Both carfilzomib and bortezomin inhibit the same proteasomal subunit (the 20S chymotrypsin-like β5 subunit), but carfilzomib does it irreversibly. In clinical practice, carfilzomib is a useful agent, because it is usually well tolerated, and does not produce significant myelosuppression and neurotoxicity [65, 66].

Histone deacetylase (HDAC) inhibitors are a class of molecules that epigenetically regulate gene transcription by modulating the structure of chromatin and DNA. These drugs can increase the transcription of genes previously downregulated by histone acetylation. The only HDAC inhibitor currently used in the treatment of MM is panobinostat, which is usually administered in combination with dexamethasone and bortezomib in patients with relapsing/refractory disease [67].

Autologous stem cell transplant (ASCT) is an important component in the treatment of MM, and is the standard of care in eligible patients, especially in patients younger than 65–70 years [46]. Patients should have a good performance status and no prohibitive comorbidities. Stem cells are mobilized from the peripheral

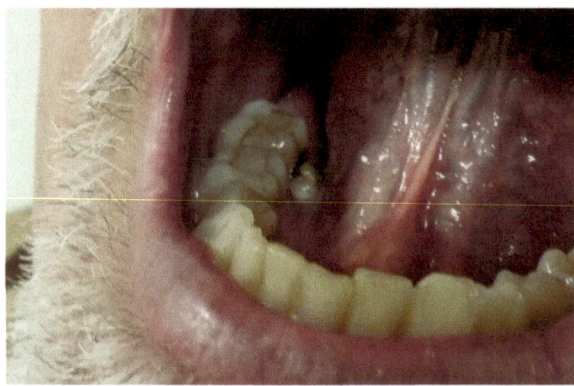

Fig. 2.7 Bisphosphonate-related osteonecrosis of the jaw involving the right mandible in a patient with multiple myeloma

blood using only growth factors (e.g., G-CSF) or chemotherapy, and collected through one or more sessions of apheresis. During the transplant process, patients receive myeloablative doses of chemotherapy (usually melphalan 200 mg/m^2), followed by the reinfusion of the blood stem cells. Autologous stem cells are cryopreserved with dimethyl sulfoxide (DMSO), and stored in liquid nitrogen for future use. In some cases, patients undergo a second autologous ASCT, either early ("tandem" ASCT), usually completed within 6 months of the first one, or later during the course of the disease ("salvage" ASCT). The benefit from a tandem ASCT is controversial due to higher cost and the lack of a survival advantage reported in some randomized trials. Most transplant centers consider second transplant a valuable option only in patients who failed to achieve CR or VGPR after the first ASCT [68]. Maintenance therapy after ASCT is a common strategy, and is usually done with lenalidomide, thalidomide, or bortezomib, however, the type and length of maintenance remains controversial [68]. In the autologous setting, the stem cells act only as "supportive therapy", because they allow the bone marrow recovery after administration of high-dose chemotherapy. In highly selected patients, *allogeneic transplantation*, from a human leukocyte antigen (HLA) matched donor, either related (e.g., sibling, or children/parents in haploidentical transplants) or unrelated (e.g., unrelated adult or umbilical cord transplant) have been performed. In this case, the stem cells are the mediators of the therapeutic effect, due to the graft-versus-myeloma effect. However, the relatively high morbidity and mortality with this approach has limited the routine use of allogeneic transplant in MM, and currently this strategy is mainly considered for younger patients with high-risk disease, preferably in the context of clinical trials [68].

In conclusion, a variety of therapies are available as options for MM. In most cases, patients receive combination regimens, which induce higher response rates and improve clinical outcomes compared with single agents. A regimen combining bortezomib, lenalidomide, and dexamethasone, can produce a response rate of 100% in patients with newly diagnosed disease [69]. The choice of a specific treatment protocol over another is determined by several factors, such as patient preference, comorbidities, and clinical behavior/disease aggressiveness [46]. Newer drugs continue to be discovered and successfully used, including carfilzomib, pomalidomide, and panobinostat, approved by the U.S. FDA in 2012, 2013, and 2015, respectively.

Monoclonal antibodies are the newest addition to the therapeutic armamentarium of MM. Two drugs have been recently introduced in the clinical practice, daratumumab and elotuzumab. Daratumumab is a human IgG1-kappa monoclonal antibody directed against CD38, a transmembrane glycoprotein expressed by myeloma cells. In heavily pretreated patients, response rate is about 35% and the median duration of response is about 7 months. Of note, administration of daratumumab can lead to two laboratory problems: (a) interference with blood typing: this drug binds to the CD38 expressed on RBCs, and patients may develop a positive Coombs test without evidence of hemolysis. This positivity can persist for up to 6 months after the daratumumab infusion. Blood banks should be notified about the use of daratumumab in a patient who needs a blood transfusion; (b) interference with response assessment: since daratumumab is a monoclonal antibody, SPEP and IFE may result positive in the absence of a MM paraprotein. This creates problems of interpretation of test results in patients with IgG MM who are in complete remission. A specific IFE reflex assay can be used to abrogate the interference. Elotuzumab is a humanized IgG1 monoclonal antibody directed against CS1 (cell-surface glycoprotein CD2 subset 1, CD319), also called SLAMF7 (Signaling Lymphocytic Activation Molecule Family 7). CS1 is highly expressed on MM cells, whereas its expression in normal tissues is absent (with the exception of normal plasma cells, NK cells, and CD8+ T-cells). Elotuzumab prevents adhesion of MM cells to the bone marrow stromal cells, it exerts a direct activation of NK cells, and induces antibody-dependent cell-mediated cytotoxicity. It does not have a significant single-agent activity, but it prolongs progression-free survival when administered in combination with lenalidomide and dexamethasone.

Radiation therapy (*RT*) has only a palliative role in MM, and is used for treating bone pain (42%), pathological fractures (28%), spinal cord compression (10%), and involvement of vital organs/extramedullary disease (10%) [70]. In solitary plasmacytoma, either osseous or extraosseous, RT is a very important modality of treatment, as it may lead to a definitive cure in a substantial number of cases.

Supportive Therapy

Supportive therapy includes measures which apply to all patients, and steps required for managing specific chemotherapy agents. In the first category, intravenous biphosphonates, i.e., pamidronate and zoledronic acid, are used for the prevention and treatment of skeletal-related complications: osteoporosis, hypercalcemia, and pathologic fractures. These compounds should be administered for at least 2 years after the initial diagnosis [71]. In a recently reported clinical trial, zoledronic acid was found to reduce mortality and improve progression-free survival as compared to clodronate, an oral bisphosphonate [72]. Importantly, based on the results of the same trial, bisphosphonates are now recommended for all MM patients receiving antineoplastic therapy, regardless of documented bone disease. A rare (about 4%) but typical complication of bisphosphonates in MM is osteonecrosis of the jaw, with bone exposure through an opening in the buccal mucosa (Fig. 2.7). Supportive measures that may be needed in all patients include: (a) transfusions and recombinant erythropoietin, to treat anemia; (b) adequate hydration and plasma exchange in case of renal insufficiency; (c) plasmapheresis for hyperviscosity syndrome; (d) antimicrobials, intravenous immunoglobulins (IVIG), and vaccinations to prevent or treat infections; (e) analgesics including opioid narcotics to treat pain; and (f) vertebral augmentation procedures (vertebroplasty and kyphoplasty), to relieve pain secondary to vertebral compression fractures. Vertebroplasty involves the percutaneous injection of polymethyl methacrylate (PMMA), a highly viscous bone cement, in the vertebral body. Kyphoplasty requires the insertion of an inflatable balloon, creating a cavity in the vertebral body that can be filled with PMMA. The introduction of inflatable bone tamps into the vertebral body can help restore the vertebral body to its original height [73].

Supportive care required with specific chemotherapy agents include: (a) aspirin and low-molecular-weight heparin, for the prevention of venous thromboembolism associated with IMiD therapy; (b) proton pump inhibitors to prevent corticosteroid-induced gastritis; and (c) antineuropathic drugs such as gabapentin or pregabalin, for treatment of the peripheral neuropathy, primarily seen after borte-zomib and thalidomide therapy.

Prognosis and Clinical Outcomes

Therapeutic advancement in recent years has improved the OS of myeloma patients. The clinical benefits have been attributed not only to the autologous ASCT, but also to the discovery of the novel agents, like the IMiDs and the proteasome inhibitors [74]. Despite the use of these new agents, and the improvement of the median survival from 3 to 6 years in the last 20 years [3], MM is generally considered an incurable disease, and relapse occurs in the vast majority of patients [45]. Several factors influence the prognosis of MM (Table 2.3) and life expectancy is highly variable, ranging from just a few months to more than a decade.

The International Staging System (ISS) provides the simplest and most powerful tool for risk stratification in MM, and has replaced the Durie-Salmon (DS) system [75]. Stage I is defined by an albumin level ≥3.5 g/dL and beta-2-microglobulin ($\beta2$ m) <3.5 mg/dL, and stage III by a $\beta2$ m > 5.5 mg/dL. Stage II includes patients not included in stage I or II. The $\beta2$ m is the light chain component of the HLA class I antigen complex synthesized by all nucleated cells. It reflects not only tumor burden, but also renal function. The serum albumin level reflects the nutri-tional status and the presence of nephrotic-range proteinuria. The ISS is a better staging system than the old Durie-Salmon (DS) system, because it is easier to determine, it distributes patients more uniformly across the three stages, and because it is not affected by interobserver variability (unlike the interpretation of the skeletal survey in the DS system). The Revised ISS (R-ISS) published in 2015 distinguishes 3 stages: stage I is defined by ISS I, normal LDH, and no high-risk CA (5-year survival 82%); stage II includes patients not in either stage I or III (5-year survival 62%); and stage III is defined by ISS III + either high LDH or high-risk cytogenetic abnormalities (5-year survival 40%). High-risk chromosomal abnormalities (CA) were defined by the presence of 17p-, t(4;14), or t(14;16), as detected by FISH.

Certain cytogenetic features are associated with high-risk MM. These include the deletion of the 17p13 locus (which contains the tumor suppressor gene p53) [76], the t(4;14) translocation [77], the 1q21 gain/amplification, and the 1p deletion [78]. The t(11;14) has a neutral impact on prognosis [79], whereas hyperdiploidy has a relatively favorable course [80]. Data for the t(14;16) translocations are conflicting. Some authors believe t(14;16) carries a negative impact on prognosis [81], whereas others did not confirm this finding [77]. More recently, several groups have identified high-risk signatures in gene expression profiling (GEP), with models including 15 genes [82], 70 genes [83], or 92 genes [84]. However, at present time GEP is not routinely used in the clinical practice.

The prognosis of AL amyloidosis is highly variable, as in MM, but the prognosis of AL amyloidosis is generally worse. Thirty percent of patients die of the disease in the first year after diagnosis [85]. The most important factor for risk stratification is the presence of cardiac involvement. In a study, patients with stage III cardiac amyloidosis (defined as levels of NT-pro-BNP and troponin T > 332 ng/L and >0.035 µg/L, respectively) had a median survival of only 3–4 months [86]. The staging system for amyloidosis has been revised, and includes four stages, depending on the levels of free light chains and cardiac biomarkers NT-pro-BNP and troponin T. Median survival ranges from 60 months in patients with stage I, to 6 months in patients with stage IV [87].

The prognosis of SP and isolated EMPC is excellent, with a median survival of greater than 10 years [88, 89].

Conclusion

Cytogenetic analysis and molecular biology studies indicate that MM is a heterogeneous disease. Its genetic complexity is not only due to the presence of several cytogenetic and molecular subtypes within a single patient, but also by the presence of intraclonal heterogeneity and progression with branching evolutionary patterns. Therefore, it is likely that in the future, management will be refined by risk-adapted approaches and individualized treatment. Resistance to therapy will likely be prevented or treated with the individualized application of genomic and proteomic analyses, targeting vulnerable pathways of a specific patient in a specific phase of the disease. Clearly, further advances in the treatment of MM, and the ultimate goal of a cure, could not be achieved by conventional chemotherapy regimens, but require the integration of novel small molecules and biologics that specifically target the pathological mechanisms underlying the progression of MM. Ultimately, gene and protein profiling and oncogenomic studies will identify target molecules which contribute to the pathogenesis and progression of MM, and will direct the development of new agents that block the molecular events that initiate and promote the growth of the disease.

References

1. Siegel R, Naishadham D, Jemal A. Cancer statistics. CA Cancer J Clin. 2013;63:11–30.
2. Blade J, Kyle RA. Multiple myeloma in young patients: clinical presentation and treatment approach. Leuk Lymphoma. 1998;30:493–501.
3. Rollig C, Knop S, Bornhauser M. Multiple myeloma. Lancet 2014.
4. Landgren O, Kyle RA, Pfeiffer RM, et al. Monoclonal gammopathy of undetermined significance (MGUS) consistently precedes multiple myeloma: a prospective study. Blood. 2009;113:5412–7.
5. Weiss BM, Abadie J, Verma P, et al. A monoclonal gammopathy precedes multiple myeloma in most patients. Blood. 2009;113:5418–22.
6. Kyle RA, Therneau TM, Rajkumar SV, et al. Prevalence of monoclonal gammopathy of undetermined significance. N Engl J Med. 2006;354:1362–9.
7. Kyle RA, Therneau TM, Rajkumar SV, et al. A long-term study of prognosis in monoclonal gammopathy of undetermined significance. N Engl J Med. 2002;346:564–9.
8. Kyle RA, Therneau TM, Rajkumar SV, et al. Long-term follow-up of IgM monoclonal gammopathy of undetermined significance. Blood. 2003;102:3759–64.
9. Kyle RA, Durie BG, Rajkumar SV, et al. Monoclonal gammopathy of undetermined significance (MGUS) and smoldering (asymptomatic) multiple myeloma: IMWG consensus perspectives risk factors for progression and guidelines for monitoring and management. Leukemia. 2010;24:1121–7.
10. Mangiacavalli S, Cocito F, Pochintesta L, et al. Monoclonal gammopathy of undetermined significance: a new proposal of workup. Eur J Haematol. 2013;91:356–60.
11. Rago A, Grammatico S, Za T, et al. Prognostic factors associated with progression of smoldering multiple myeloma to symptomatic form. Cancer. 2012;118:5544–9.
12. Larsen JT, Kumar SK, Dispenzieri A, et al. Serum free light chain ratio as a biomarker for high-risk smoldering multiple myeloma. Leukemia. 2013;27:941–6.
13. Rajkumar SV, Dimopoulos MA, Palumbo A, et al. International Myeloma Working Group updated criteria for the diagnosis of multiple myeloma. Lancet Oncol. 2014;15:e538–48.
14. Kakkar N, Das S. Relapse of multiple myeloma: diagnosis by clot section alone with negative bone marrow aspirate and trephine biopsy. Indian J Pathol Microbiol. 2009;52:290–1.
15. Ceneli O, Haznedar R. Advanced multiple myeloma with negative bone marrow biopsy and positive soft tissue lesions in the (18)F-FDG PET/CT scan. Hell J Nucl Med. 2008;11:56–7.
16. Roodman GD. Mechanisms of bone metastasis. N Engl J Med. 2004;350:1655–64.
17. Start DA, Silva FG, Davis LD, et al. Myeloma cast nephropathy: immunohistochemical and lectin studies. Mod Pathol. 1988;1:336–47.
18. Ying WZ, Sanders PW. Mapping the binding domain of immunoglobulin light chains for Tamm-Horsfall protein. Am J Pathol. 2001;158:1859–66.
19. Maldonado JE, Velosa JA, Kyle RA, et al. Fanconi syndrome in adults. A manifestation of a latent form of myeloma. Am J Med. 1975;58:354–64.

20. Papanikolaou X, Repousis P, Tzenou T, et al. Incidence, clinical features, laboratory findings and outcome of patients with multiple myeloma presenting with extramedullary relapse. Leuk Lymphoma. 2013;54:1459–64.
21. Pour L, Sevcikova S, Greslikova H, et al. Soft-tissue extramedullary multiple myeloma prognosis is significantly worse in comparison to bone-related extramedullary relapse. Haematologica. 2014;99:360–4.
22. Talamo G, Dolloff NG, Sharma K, et al. Clinical features and outcomes of plasma cell leukemia: a single-institution experience in the era of novel agents. Rare Tumors. 2012;4:e39.
23. Pagano L, Valentini CG, De Stefano V, et al. Primary plasma cell leukemia: a retrospective multicenter study of 73 patients. Ann Oncol. 2011;22:1628–35.
24. Fernandez de Larrea C, Kyle RA, Durie BG, et al. Plasma cell leukemia: consensus statement on diagnostic requirements, response criteria and treatment recommendations by the International Myeloma Working Group. Leukemia. 2013;27:780–91.
25. Pozzi C, D'Amico M, Fogazzi GB, et al. Light chain deposition disease with renal involvement: clinical characteristics and prognostic factors. Am J Kidney Dis. 2003;42:1154–63.
26. Dispenzieri A. POEMS syndrome: 2011 update on diagnosis, risk-stratification, and management. Am J Hematol. 2011;86:591–601.
27. Palumbo A, Rajkumar SV, San Miguel JF, et al. International Myeloma Working Group consensus statement for the management, treatment, and supportive care of patients with myeloma not eligible for standard autologous stem-cell transplantation. J Clin Oncol. 2014;32:587–600.
28. Katzmann JA, Clark R, Wiegert E, et al. Identification of monoclonal proteins in serum: a quantitative comparison of acetate, agarose gel, and capillary electrophoresis. Electrophoresis. 1997;18:1775–80.
29. Bienvenu J, Graziani MS, Arpin F, et al. Multicenter evaluation of the Paragon CZE 2000 capillary zone electrophoresis system for serum protein electrophoresis and monoclonal component typing. Clin Chem. 1998;44:599–605.
30. Dispenzieri A, Kyle R, Merlini G, et al. International Myeloma Working Group guidelines for serum-free light chain analysis in multiple myeloma and related disorders. Leukemia. 2009;23:215–24.
31. Wozney JL, Damluji AA, Ahmed F, et al. Estimation of daily proteinuria in patients with multiple myeloma by using the protein-to-creatinine ratio in random urine samples. Acta Haematol. 2010;123:226–9.
32. Katzmann JA, Kyle RA, Benson J, et al. Screening panels for detection of monoclonal gammopathies. Clin Chem. 2009;55:1517–22.
33. Durie BG, Harousseau JL, Miguel JS, et al. International uniform response criteria for multiple myeloma. Leukemia. 2006;20:1467–73.
34. San Miguel JF, Sanchez J, Gonzalez M. Prognostic factors and classification in multiple myeloma. Br J Cancer. 1989;59:113–8.
35. Bartl R, Frisch B, Fateh-Moghadam A, et al. Histologic classification and staging of multiple myeloma. A retrospective and prospective study of 674 cases. Am J Clin Pathol. 1987;87:342–55.
36. Seegmiller AC, Xu Y, McKenna RW, et al. Immunophenotypic differentiation between neoplastic plasma cells in mature B-cell lymphoma vs plasma cell myeloma. Am J Clin Pathol. 2007;127:176–81.
37. Paiva B, Martinez-Lopez J, Vidriales MB, et al. Comparison of immunofixation, serum free light chain, and immunophenotyping for response evaluation and prognostication in multiple myeloma. J Clin Oncol. 2011;29:1627–33.
38. Rawstron AC, Davies FE, DasGupta R, et al. Flow cytometric disease monitoring in multiple myeloma: the relationship between normal and neoplastic plasma cells predicts outcome after transplantation. Blood. 2002;100:3095–100.

39. Dimopoulos M, Terpos E, Comenzo RL, et al. International myeloma working group consensus statement and guidelines regarding the current role of imaging techniques in the diagnosis and monitoring of multiple Myeloma. Leukemia. 2009;23:1545–56.
40. Dimopoulos MA, Hillengass J, Usmani S, et al. Role of magnetic resonance imaging in the management of patients with multiple myeloma: a consensus statement. J Clin Oncol. 2015;33:657–64.
41. Zamagni E, Patriarca F, Nanni C, et al. Prognostic relevance of 18-F FDG PET/CT in newly diagnosed multiple myeloma patients treated with up-front autologous transplantation. Blood. 2011;118:5989–95.
42. Nanni C, Zamagni E, Celli M, et al. The value of 18F-FDG PET/CT after autologous stem cell transplantation (ASCT) in patients affected by multiple myeloma (MM): experience with 77 patients. Clin Nucl Med. 2013;38:e74–9.
43. Abildgaard N, Brixen K, Kristensen JE, et al. Assessment of bone involvement in patients with multiple myeloma using bone densitometry. Eur J Haematol. 1996;57:370–6.
44. Berenson JR, Rosen LS, Howell A, et al. Zoledronic acid reduces skeletal-related events in patients with osteolytic metastases. Cancer. 2001;91:1191–200.
45. Kumar SK, Rajkumar SV, Dispenzieri A, et al. Improved survival in multiple myeloma and the impact of novel therapies. Blood. 2008;111:2516–20.
46. Reed V, Shah J, Medeiros LJ, et al. Solitary plasmacytomas: outcome and prognostic factors after definitive radiation therapy. Cancer. 2011;117:4468–74.
47. Gerry D, Lentsch EJ. Epidemiologic evidence of superior outcomes for extramedullary plasmacytoma of the head and neck. Otolaryngol Head Neck Surg. 2013;148:974–81.
48. Mateos MV, Hernandez MT, Giraldo P, et al. Lenalidomide plus dexamethasone for high-risk smoldering multiple myeloma. N Engl J Med. 2013;369:438–47.
49. Fermand JP, Ravaud P, Chevret S, et al. High-dose therapy and autologous peripheral blood stem cell transplantation in multiple myeloma: up-front or rescue treatment? Results of a multicenter sequential randomized clinical trial. Blood. 1998;92:3131–6.
50. Kumar SK, Lacy MQ, Dispenzieri A, et al. Early versus delayed autologous transplantation after immunomodulatory agents-based induction therapy in patients with newly diagnosed multiple myeloma. Cancer. 2012;118:1585–92.
51. Kumar SK, Therneau TM, Gertz MA, et al. Clinical course of patients with relapsed multiple myeloma. Mayo Clin Proc. 2004;79:867–74.
52. Kumar S, Lacy MQ, Dispenzieri A, et al. Single agent dexamethasone for pre-stem cell transplant induction therapy for multiple myeloma. Bone Marrow Transplant. 2004;34:485–90.
53. Jagannath S, Kyle RA, Palumbo A, et al. The current status and future of multiple myeloma in the clinic. Clin Lymphoma Myeloma Leuk. 2010;10:28–43.
54. Zhu YX, Braggio E, Shi CX, et al. Cereblon expression is required for the antimyeloma activity of lenalidomide and pomalidomide. Blood. 2011;118:4771–9.
55. Kronke J, Udeshi ND, Narla A, et al. Lenalidomide causes selective degradation of IKZF1 and IKZF3 in multiple myeloma cells. Science. 2014;343:301–5.
56. Lu G, Middleton RE, Sun H, et al. The myeloma drug lenalidomide promotes the cereblon-dependent destruction of Ikaros proteins. Science. 2014;343:305–9.
57. Palumbo A, Facon T, Sonneveld P, et al. Thalidomide for treatment of multiple myeloma: 10 years later. Blood. 2008;111:3968–77.
58. Singhal S, Mehta J, Desikan R, et al. Antitumor activity of thalidomide in refractory multiple myeloma. N Engl J Med. 1999;341:1565–71.
59. Weber DM, Chen C, Niesvizky R, et al. Lenalidomide plus dexamethasone for relapsed multiple myeloma in North America. N Engl J Med. 2007;357:2133–42.
60. Lacy MQ, Hayman SR, Gertz MA, et al. Pomalidomide (CC4047) plus low dose dexamethasone (Pom/dex) is active and well tolerated in lenalidomide refractory multiple myeloma (MM). Leukemia. 2010;24:1934–9.
61. Obeng EA, Carlson LM, Gutman DM, et al. Proteasome inhibitors induce a terminal unfolded protein response in multiple myeloma cells. Blood. 2006;107:4907–16.

62. Richardson PG, Barlogie B, Berenson J, et al. A phase 2 study of bortezomib in relapsed, refractory myeloma. N Engl J Med. 2003;348:2609–17.
63. Richardson PG, Sonneveld P, Schuster MW, et al. Bortezomib or high-dose dexamethasone for relapsed multiple myeloma. N Engl J Med. 2005;352:2487–98.
64. O'Connor OA, Stewart AK, Vallone M, et al. A phase 1 dose escalation study of the safety and pharmacokinetics of the novel proteasome inhibitor carfilzomib (PR-171) in patients with hematologic malignancies. Clin Cancer Res. 2009;15:7085–91.
65. Jagannath S, Vij R, Stewart AK, et al. An open-label single-arm pilot phase II study (PX-171-003-A0) of low-dose, single-agent carfilzomib in patients with relapsed and refractory multiple myeloma. Clin Lymphoma Myeloma Leuk. 2012;12:310–8.
66. Siegel DS, Martin T, Wang M, et al. A phase 2 study of single-agent carfilzomib (PX-171-003-A1) in patients with relapsed and refractory multiple myeloma. Blood. 2012;120:2817–25.
67. San-Miguel JF, Hungria VT, Yoon SS, et al. Panobinostat plus bortezomib and dexamethasone versus placebo plus bortezomib and dexamethasone in patients with relapsed or relapsed and refractory multiple myeloma: a multicentre, randomised, double-blind phase 3 trial. Lancet Oncol. 2014;15:1195–206.
68. Shah N, Callander N, Ganguly S, et al. Hematopoietic Stem Cell Transplantation for Multiple Myeloma: Guidelines from the American Society for Blood and Marrow Transplantation. Biol Blood Marrow Transplant 2015.
69. Richardson PG, Weller E, Lonial S, et al. Lenalidomide, bortezomib, and dexamethasone combination therapy in patients with newly diagnosed multiple myeloma. Blood. 2010;116:679–86.
70. Talamo G, Dimaio C, Abbi KK, et al. Current role of radiation therapy for multiple myeloma. Front Oncol. 2015;5:40.
71. Terpos E, Morgan G, Dimopoulos MA, et al. International Myeloma Working Group recommendations for the treatment of multiple myeloma-related bone disease. J Clin Oncol. 2013;31:2347–57.
72. Morgan GJ, Davies FE, Gregory WM, et al. First-line treatment with zoledronic acid as compared with clodronic acid in multiple myeloma (MRC Myeloma IX): a randomised controlled trial. Lancet. 2010;376:1989–99.
73. Berenson J, Pflugmacher R, Jarzem P, et al. Balloon kyphoplasty versus non-surgical fracture management for treatment of painful vertebral body compression fractures in patients with cancer: a multicentre, randomised controlled trial. Lancet Oncol. 2011;12:225–35.
74. Richardson PG, Mitsiades C, Schlossman R, et al. New drugs for myeloma. Oncologist. 2007;12:664–89.
75. Greipp PR, San Miguel J, Durie BG, et al. International staging system for multiple myeloma. J Clin Oncol. 2005;23:3412–20.
76. Avet-Loiseau H, Attal M, Moreau P, et al. Genetic abnormalities and survival in multiple myeloma: the experience of the Intergroupe Francophone du Myelome. Blood. 2007;109:3489–95.
77. Avet-Loiseau H, Malard F, Campion L, et al. Translocation t(14;16) and multiple myeloma: is it really an independent prognostic factor? Blood. 2011;117:2009–11.
78. Carrasco DR, Tonon G, Huang Y, et al. High-resolution genomic profiles define distinct clinico-pathogenetic subgroups of multiple myeloma patients. Cancer Cell. 2006;9:313–25.
79. Fonseca R, Barlogie B, Bataille R, et al. Genetics and cytogenetics of multiple myeloma: a workshop report. Cancer Res. 2004;64:1546–58.
80. Pawlyn C, Melchor L, Murison A, et al. Coexistent hyperdiploidy does not abrogate poor prognosis in myeloma with adverse cytogenetics and may precede IGH translocations. Blood. 2015;125:831–40.
81. Fonseca R, Blood E, Rue M, et al. Clinical and biologic implications of recurrent genomic aberrations in myeloma. Blood. 2003;101:4569–75.
82. Decaux O, Lode L, Magrangeas F, et al. Prediction of survival in multiple myeloma based on gene expression profiles reveals cell cycle and chromosomal instability signatures in high-risk

patients and hyperdiploid signatures in low-risk patients: a study of the Intergroupe Francophone du Myelome. J Clin Oncol. 2008;26:4798–805.

83. Shaughnessy JD Jr, Zhan F, Burington BE, et al. A validated gene expression model of high-risk multiple myeloma is defined by deregulated expression of genes mapping to chromosome 1. Blood. 2007;109:2276–84.

84. Kuiper R, Broyl A, de Knegt Y, et al. A gene expression signature for high-risk multiple myeloma. Leukemia. 2012;26:2406–13.

85. Kumar SK, Gertz MA, Lacy MQ, et al. Recent improvements in survival in primary systemic amyloidosis and the importance of an early mortality risk score. Mayo Clin Proc. 2011;86:12–8.

86. Dispenzieri A, Gertz MA, Kyle RA, et al. Prognostication of survival using cardiac troponins and N-terminal pro-brain natriuretic peptide in patients with primary systemic amyloidosis undergoing peripheral blood stem cell transplantation. Blood. 2004;104:1881–7.

87. Kumar S, Dispenzieri A, Lacy MQ, et al. Revised prognostic staging system for light chain amyloidosis incorporating cardiac biomarkers and serum free light chain measurements. J Clin Oncol. 2012;30:989–95.

88. Knobel D, Zouhair A, Tsang RW, et al. Prognostic factors in solitary plasmacytoma of the bone: a multicenter Rare Cancer Network study. BMC Cancer. 2006;6:118.

89. Bachar G, Goldstein D, Brown D, et al. Solitary extramedullary plasmacytoma of the head and neck–long-term outcome analysis of 68 cases. Head Neck. 2008;30:1012–9.

90. Talamo G, Farooq U, Zangari M, et al. Beyond the CRAB symptoms: a study of presenting clinical manifestations of multiple myeloma. Clin Lymphoma Myeloma Leuk. 2010;10:464–8.

Chapter 3
Morphologic and Immunohistochemical Evaluation of Plasma Cell Neoplasms

Soumya Pandey and Robert B. Lorsbach

Introduction

Plasma cell myeloma (PCM) is a bone marrow-based neoplasm, the diagnosis of which is straightforward in most cases. The presence of marked cytologic pleomorphism and extramedullary involvement, for example, can make the diagnosis of myeloma a difficult one in some instances. This chapter will provide a comprehensive review of the morphologic features of PCM; however, our emphasis will be on those variant histologic, cytologic, and immunophenotypic features that provide important diagnostic clues or that have the potential for misinterpretation leading to possible misdiagnosis. Because it is predominantly a neoplasm of the bone marrow, our discussion will focus primarily on bone marrow findings in PCM. However, we will also discuss the morphologic features of peripheral blood (i.e., plasma cell leukemia) and extramedullary involvement, where PCM must be distinguished from solid tumors and other hematolymphoid malignancies.

S. Pandey (✉)
Department of Pathology, University of Arkansas for Medical Sciences, Little Rock, AR, USA
e-mail: spandey@uams.edu

R.B. Lorsbach
Department of Pathology and Laboratory Medicine, Cincinnati Children's Hospital Medical Center, University of Cincinnati College of Medicine, 3333 Burnet Ave., ML 1035, Cincinnati, OH 45229, USA
e-mail: robert.lorsbach@cchmc.org

© Springer International Publishing Switzerland 2016
R.B. Lorsbach and M. Yared (eds.), *Plasma Cell Neoplasms*,
DOI 10.1007/978-3-319-42370-8_3

Normal Bone Marrow Plasma Cells: General Findings and Morphologic Features

Plasma cells (PCs) are normal residents of the bone marrow and usually comprise a small minority of the overall cellularity. PCs are rare in pediatric bone marrow, especially in young children, where they usually account for less than 1% of bone marrow cells. They are present at a somewhat higher frequency in adult bone marrow, but normally still comprise less than 5% of the cellularity. Reactive bone marrow plasmacytosis is commonly seen in both the adult and pediatric settings and is associated with a wide array of neoplastic, infectious, and inflammatory disorders (Table 3.1). In some instances, most notably in the setting of HIV/AIDS, bone marrow plasmacytosis may be so striking as to mimic PCM.

In the bone marrow aspirate, reactive PCs have a mature morphology (sometimes referred to as Marshalko morphology) characterized cytologically by small size, dense clumped chromatin, absent or inconspicuous nucleoli, and relatively abundant basophilic cytoplasm in which the nucleus is often eccentrically positioned (Box 3.1).

Box 3.1: Bone marrow findings in reactive plasmacytosis

Marrow localization

> Perivascular or small interstitial clusters
> No effacement of normal marrow architecture

Cytologic features

> Small cell size (8–15 µM diameter)
> Mature cytology (round nucleus, clumped chromatin, indistinct nucleoli, eccentric cuff of basophilic cytoplasm)
> Minimal binucleation
> Few cytoplasmic inclusions/vacuoles

Unlike other lymphoid cells, normal PCs manifest very low-level binucleation. The presence of multinucleated or frequent binucleated PCs is uncommon in reactive settings and should raise the suspicion for PCM. Although reactive

Table 3.1 Causes of reactive bone marrow plasmacytosis

Infectious	Inflammatory/autoimmune disorders	Malignancy	Post-chemotherapy
HIV/AIDS	Systemic lupus erythrematosus	Hodgkin lymphoma	–
Epstein-Barr viral infection	Sjogren's syndrome	Angioimmunoblastic T-cell lymphoma	–

plasmacytoses are comprised of mature-appearing PCs, it should be kept in mind that a subset of PC neoplasms, in particular monoclonal gammopathy of undetermined significance (MGUS), is comprised of Marshalko-type plasma cells that are morphologically indistinguishable from their benign counterparts.

Normal PCs reside in the marrow as single cells or in small clusters usually within perivascular niches or scattered within the interstitium (Box 3.1), features best appreciated in the bone marrow biopsy and highlighted by immunohistochemistry (IHC) for plasmacytic antigens, e.g., CD138 (Fig. 3.1). Prominent reactive plasmacytoses are relatively common as discussed above. There are diagnostic clues to the reactive nature of even the most florid reactive plasmacytosis in addition to the aforementioned cytologic features. These include the preserved perivascular/interstitial localization of the PCs within the bone marrow and the absence of any significant degree of marrow architectural effacement, a common finding in PCM.

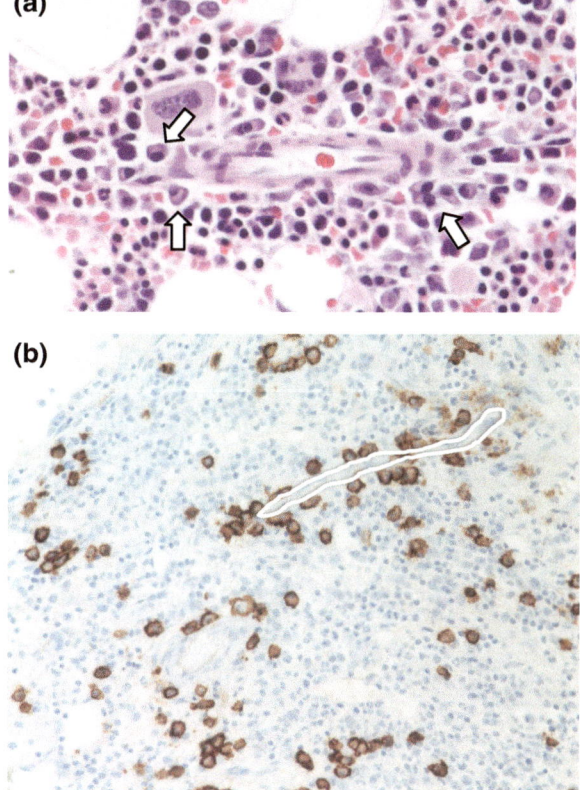

Fig. 3.1 Distribution of bone marrow plasma cells (PCs). **a** Within the bone marrow, PCs reside within perivascular niches (arrows) or as small interstitial cell aggregates. **b** This distribution of normal PCs is highlighted by CD138 immunohistochemistry, where several PCs are present cuffing a bone marrow blood vessel (demarcated by a white line)

Bone Marrow Evaluation in Plasma Cell Myeloma: Diagnostic Considerations and Practical Issues

Diagnostic criteria for MGUS, asymptomatic PCM, and symptomatic PCM are delineated in the current 2008 WHO classification (Table 3.2) [1]. These criteria are straightforward and require for final diagnosis the integration of bone marrow findings with clinical data and the results of radiologic imaging, serum protein studies, and other laboratory studies. It should be emphasized that no minimal level of monotypic PCs is specified by the WHO classification for a diagnosis of symptomatic myeloma. This reflects the fact that neoplastic PCs are present at a very low level in the initial diagnostic bone marrow specimen in a small minority of patients with otherwise typical symptomatic myeloma. Of note, the frequency of clonal bone marrow PCs is similarly low in primary amyloidosis. In the setting of low-level clonal plasmacytosis, distinction of symptomatic myeloma from primary amyloidosis requires correlation with laboratory and radiology studies; patients with organ dysfunction due solely to amyloid deposition and no other evidence of typical myelomatous end-organ damage (e.g., anemia, osteolytic lesions, hypercalcemia) should be diagnosed with primary amyloidosis rather than symptomatic myeloma.

Even with the availability of sophisticated serum protein analytic techniques, flow cytometry, and other advanced genetic analytic tests, morphologic evaluation of the bone marrow remains the critical starting point for diagnosis of the PC neoplasms. Thorough morphologic assessment should guide the judicious utilization of ancillary testing in the evaluation of diagnostic and follow-up marrow specimens. For patients in whom a PC neoplasm is suspected, evaluation of both a high-quality bone marrow aspirate and biopsy is the optimal approach for appropriate diagnosis and accurate assessment of the extent of bone marrow involvement. Indeed the aspirate and biopsy serve complimentary roles in the diagnostic workup of PC tumors (Table 3.3). There are no universally accepted criteria for bone marrow biopsy adequacy; nevertheless, a biopsy containing a minimum of 1 cm of intact bone marrow is a reasonable criterion for adequacy. To ensure adequate bone marrow sampling, aspirate smears should contain ample spicules and be comparably cellular to the accompanying bone marrow biopsy. Aspirate smears lacking

Table 3.2 WHO diagnostic criteria for plasma cell neoplasms

	M-protein (gm/dL)	Bone marrow plasma cells	Myeloma-related end-organ damage	Other
MGUS	<3 g/dL	<10%	None	No evidence of a B-cell LPD
Myeloma: Asymptomatic	>3 g/dL	>10%	None	–
Symptomatic	Present[a]	Present[b]	Yes	–

[a]Presence alone of a serum or urine M-protein is required; no minimal level is specified
[b]Demonstration of monoclonal plasma cells alone is required; no minimal threshold for bone marrow plasmacytosis is specified

Table 3.3 Complementary roles of bone marrow aspirate and biopsy in diagnostic evaluation of plasma cell neoplasms

Bone marrow aspirate	Bone marrow biopsy
Evaluation of PC cytology	More accurate quantitation of disease involvement
Amenable to flow cytometry, for optimal immunophenotypic evaluation of neoplastic PCs and to exclude coexisting B-cell LPDs	Optimal evaluation for unusual myelomas -PCM with myelofibrosis -Anaplastic PCM
	Amenable to immunohistochemistry
Amenable to genetic and molecular studies	Evaluation for coexisting conditions (e.g., amyloidosis)

spicules or showing significant hemodilution should be qualified as such in the hematopathology report.

Unlike most other hematologic malignancies, heterogeneous bone marrow involvement commonly occurs in PCM. Thus, only partial or very focal involvement may be evident in the bone marrow biopsy. Such focal myeloma involvement, particularly with poorly differentiated or anaplastic tumors, may closely mimic a metastatic carcinoma or solid tumor. In such instances, the availability of a high-quality biopsy and aspirate containing ample evaluable marrow may be critical for diagnosis. This peculiar biologic behavior of PCM manifests clinically in myeloma patients as so-called macrofocal disease, which is characterized by large lytic lesions containing abundant neoplastic plasma cells and no radiologically overt disease in the intervening marrow [2, 3]. Such biologic behavior should always be kept in mind, as even repeat random iliac crest bone marrows from patients with macrofocal PCM may show little or no morphologic evidence of myeloma despite radiologic evidence of extensive osteolytic disease. Finally, on a practical note, such heterogeneity and multifocality may lead in some instances to significant discrepancies in the extent of bone marrow myeloma involvement between the aspirate and core biopsy (Box 3.2).

Box 3.2: Plasma cell myeloma bone marrow evaluation—key differential diagnoses, caveats, and practical considerations

- Bone marrow involvement by PCM is often heterogeneous
- Patients with extensive "macrofocal" PCM may show little or no morphologic evidence of disease in random iliac crest biopsies/aspirates
- PCM can present as extramedullary or para-osseous lesions, thus mimicking lymphoma or solid tumor
- Small lymphocyte-like PCM must be distinguished from low-grade B-cell lymphoma with plasmacytic differentiation, including lymphoplasmacytic lymphoma and marginal zone lymphoma
- Anaplastic or poorly differentiated PCM must be distinguished from metastatic tumor, including carcinoma and melanoma

Plasma Cell Myeloma: Diagnostically Helpful Morphologic Findings

PCM manifests a significant degree of morphologic heterogeneity both between individual cases and sometimes prominently within a given case. Familiarity with the spectrum of cytologic findings that may be encountered in PCM is particularly helpful in the evaluation of tumors where plasmacytic differentiation is less overt, in post-therapy bone marrow specimens containing a lower frequency of neoplastic PCs, or biopsies taken from extramedullary sites for which the differential diagnosis may include non-hematolymphoid malignancies. Fortunately, myeloma cells usually manifest one or more atypical cytologic features that aid in their distinction from their benign counterparts in many cases (Box 3.3).

Box 3.3: Diagnostically useful cytologic features that aid in the distinction of neoplastic myeloma cells from benign/reactive plasma cells

Enlarged cell size
Lymphoplasmacytic morphology

- May mimic lymphoplasmacytic lymphoma
- Strongly linked to t(11;14)

Vesicular or immature chromatin
Presence of distinct or prominent nucleoli
"Clonal" cytoplasmic or nuclear inclusions
Multinucleation

Neoplastic PCs are often much larger than their benign counterparts. Whereas normal PCs are 8–15 μM in diameter, the neoplastic cells in many myelomas are significantly enlarged and may be 30–40 μM in diameter (Fig. 3.2). Very large neoplastic cells ranging in size up to 150 μM in diameter may predominate in rare cases. While aberrant size is common, it should be kept in mind that some myelomas are comprised of essentially normal-sized neoplastic cells with Marshalko morphology and can be very difficult to distinguish from benign PCs on an individual cell basis (Fig. 3.3). Similarly, the neoplastic PCs in cases of so-called small lymphocyte-like myeloma (Fig. 3.11) may be very difficult to distinguish cytologically from benign lymphocytes in the setting of low-level involvement or small B-cell lymphomas in more heavily involved marrows.

Neoplastic PCs frequently have atypical nuclear features. The chromatin of benign PCs has a characteristic "clockface" appearance due to the discontinuous distribution of dense heterochromatin around the periphery of the nuclear envelope. This typical chromatin pattern is frequently absent in neoplastic PCs, where the

Fig. 3.2 Plasma cell myeloma with prominent cytologic heterogeneity. **a** and **b** A small subset of the neoplastic PCs has mature morphology and size approximating that of benign PCs (*open arrows*). Many neoplastic PCs are significantly larger, with enlarged nuclei and more copious cytoplasm (*cross-hatched arrows*). A small number of even larger bi- or multinucleated cells are present (*filled arrows*)

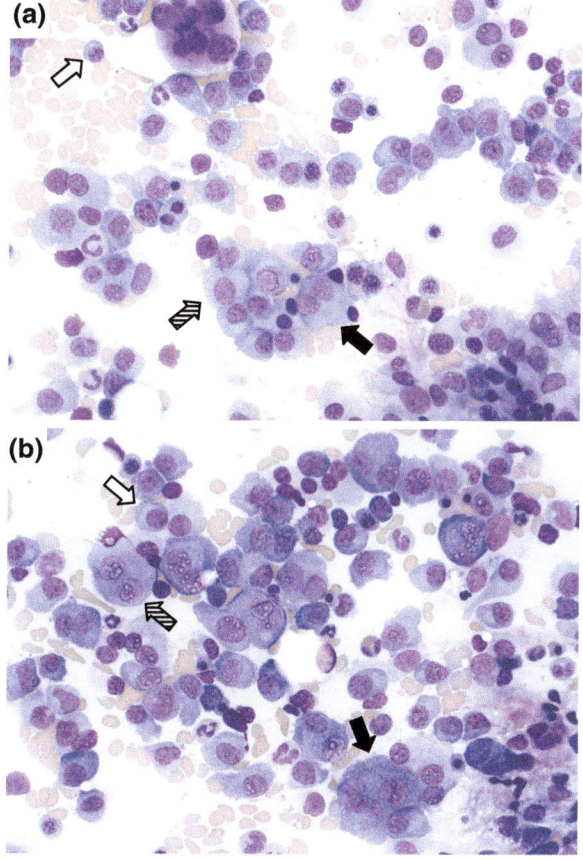

Fig. 3.3 Plasma cell myeloma with mature or Marshalko morphology. In such cases, the neoplastic PCs are virtually indistinguishable on a cytologic basis from normal bone marrow PCs

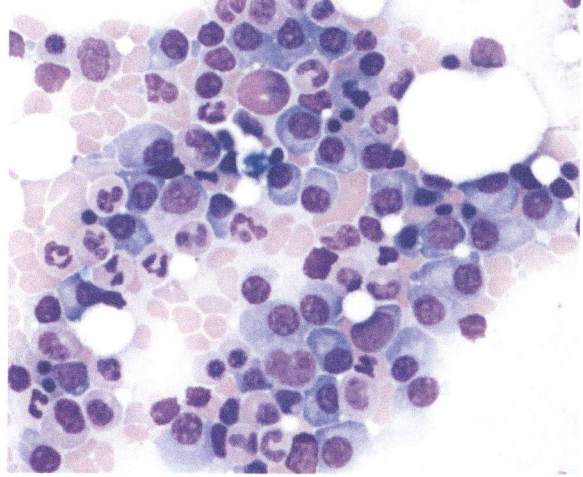

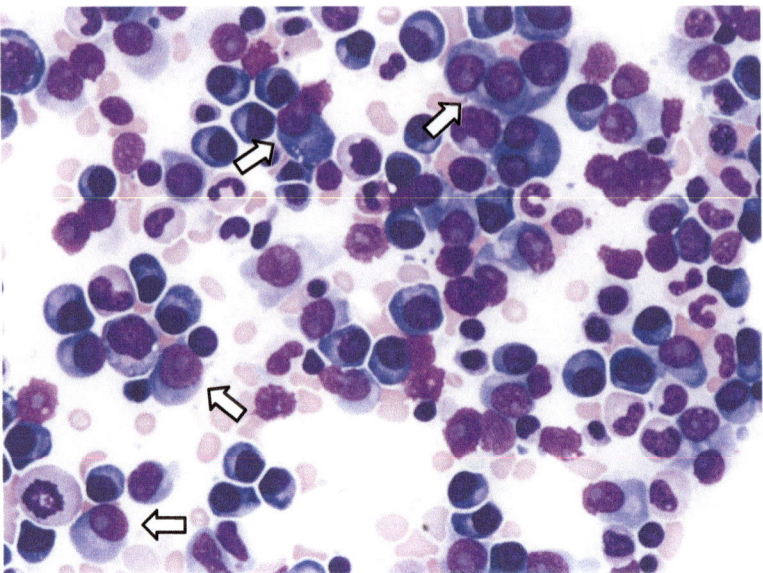

Fig. 3.4 Intermediate-grade plasma cell myeloma. While some mature-appearing cells are present, many of the neoplastic cells are enlarged, contain distinct nucleoli, and have vesicular chromatin (*arrows*)

chromatin may be vesicular or even finely dispersed and blast-like. Myeloma cells often contain distinct or even large inclusion-like nucleoli, in contrast to the inconspicuous nucleoli typical of normal PCs (Figs. 3.4 and 3.5). Such high-grade myelomas can closely mimic large cell lymphoma or even acute myeloid leukemia. Marked anaplasia is occasionally a prominent cytologic feature in PCM (Fig. 3.6) [4–8]. In addition to their large size, the myeloma cells in such tumors have prominent multinucleation or nuclear lobation, which when present in low numbers may be mistaken for megakaryocytes. Such anaplastic myelomas can pose a diagnostic challenge given their resemblance to other poorly differentiated tumors, including anaplastic large cell lymphoma, carcinoma, and melanoma. Binucleated PCs occur at a low frequency in reactive conditions. The presence of frequent binucleated or even multinucleated PCs is unusual in reactive conditions and should raise suspicion for malignancy, as binucleation is a prominent cytologic feature in some myelomas.

Cytoplasmic or nuclear inclusions are a common cytologic feature in myeloma, typically resulting from either impaired immunoglobulin secretion or synthesis of a misfolded immunoglobulin protein. These inclusions most commonly take the form of clear cytoplasmic vacuoles or eosinophilic globules, the latter known as Russell bodies (Fig. 3.7a, b); cells containing multiple Russell bodies are known as Mott

Fig. 3.5 High-grade plasma cell myeloma. **a** The neoplastic cells show only subtle features of plasmacytic differentiation. Note the frequent mitotic figures, including atypical ones. On a morphologic basis alone, such cases raise a differential diagnosis that includes large cell lymphoma and metastatic malignancies such as melanoma and carcinoma. This case was strongly CD138 positive. **b** Approximately 90% of the neoplastic myeloma cells stain positively for Ki-67. **c** Cytologic features of a high-grade PCM

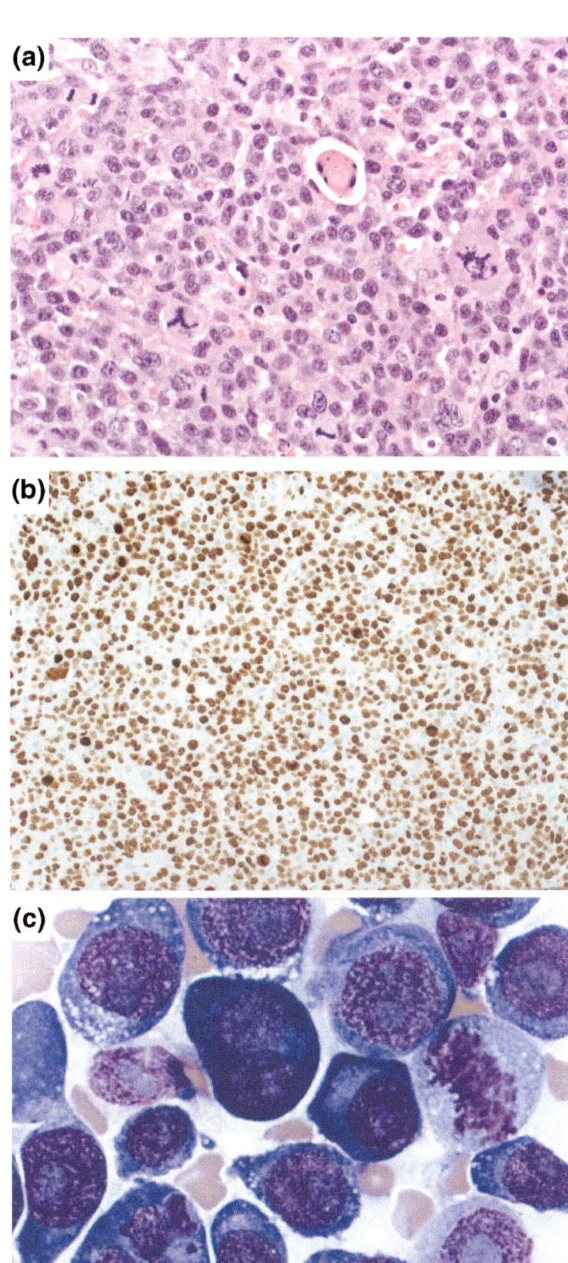

Fig. 3.6 Plasma cell myeloma with anaplastic cytology. **a** and **b** Large PCs with complex, multilobated nuclei are present in some PCM; note that these cells approximate the size of megakaryocytes. **b** Anaplastic PCs may predominate in some tumors, mimicking anaplastic large cell lymphoma or metastatic solid tumors

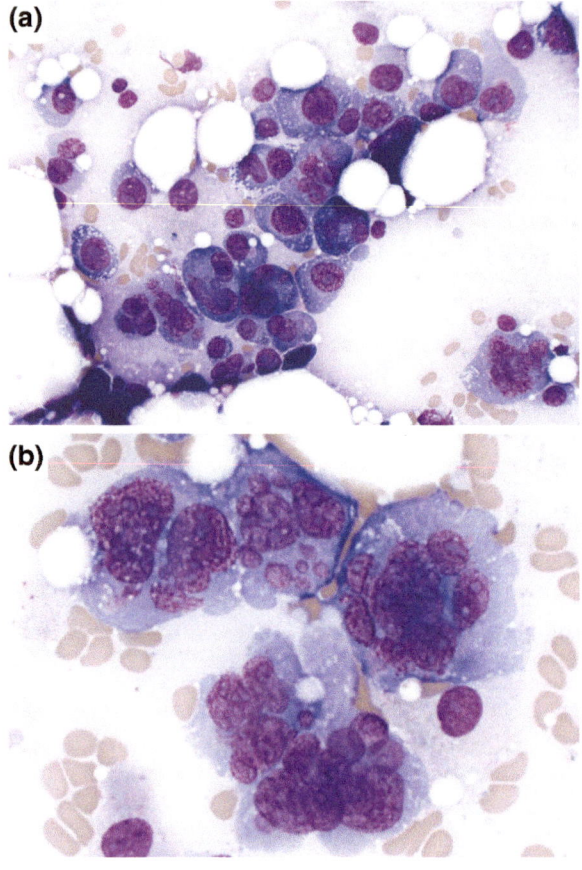

cells. Rare myelomas have signet ring cell morphology due to the presence of large optically clear vacuoles in the neoplastic cells (Fig. 3.7c, d) [9–12]. Prominent needle-like immunoglobulin inclusions resembling Auer rods are an unusual cytologic feature (Fig. 3.7e) [13, 14]. In rare cases, the cytoplasm of the neoplastic PCs has a fibrillary-like quality reminiscent of that seen in the storage histiocytes of Gaucher disease (Fig. 3.7f) [15]. Intranuclear inclusions, or so-called Dutcher bodies, are present in a subset of myelomas and are occasionally a prominent cytologic feature (Fig. 3.8a, b). Dutcher bodies are not genuine nuclear inclusions per se, but rather result from invagination of the cytoplasm into the nucleus proper [16, 17]. An association of Dutcher bodies with the presence of the t(4;14) and IgA paraprotein has been reported [18]. When present in a high fraction of cells, these inclusions may facilitate the identification of low numbers of neoplastic PCs in the bone marrow aspirate smear. With the exception of Russell bodies, most of these inclusions are present infrequently in normal bone marrow PCs.

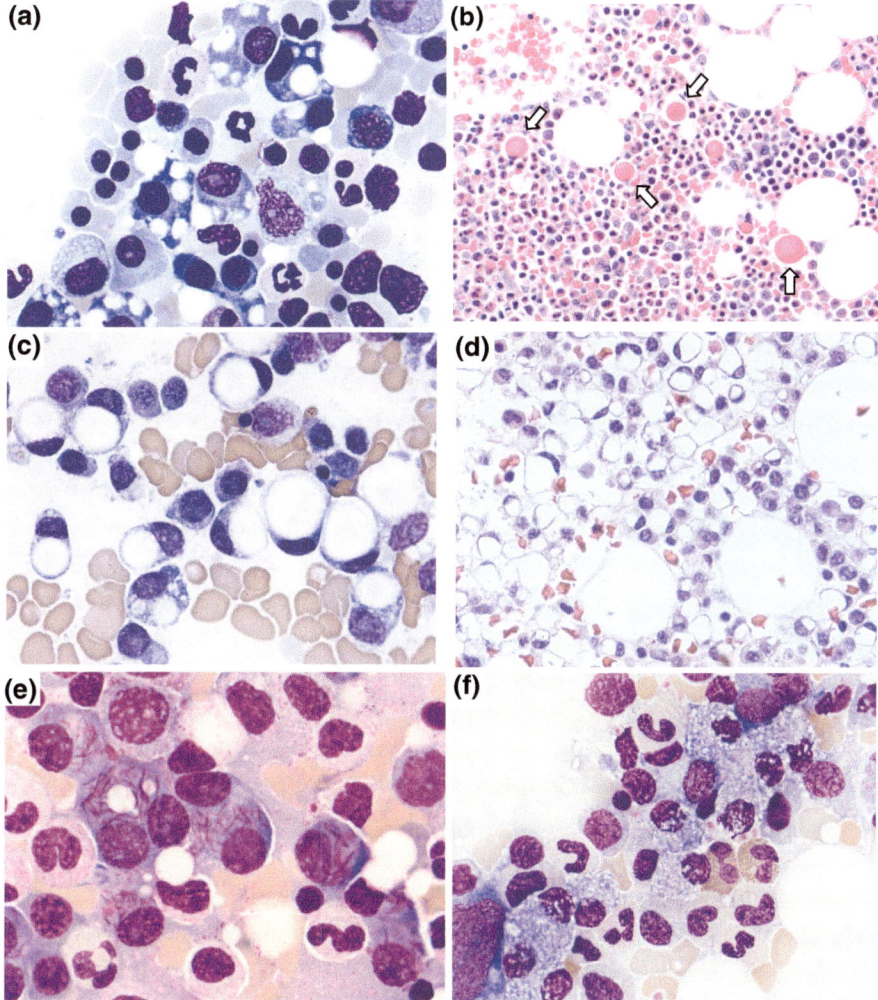

Fig. 3.7 Cytoplasmic inclusions in plasma cell myeloma. **a** Optically clear cytoplasmic vacuoles are a common finding in PCM. **b** Myeloma with prominent Russell bodies (*arrows*). Because of the size of the Russell bodies, there is distortion of the myeloma cells obscuring their cytologic features. **c** and **d** PCM with signet ring cell cytology. Cells in this rare variant are characterized by a single large vacuole that distends the cell (**c**, bone marrow aspirate; **d** bone marrow biopsy). When prominent, as in this case, signet ring cell cytology may obscure the plasmacytic nature of the neoplastic myeloma cells which may be mistaken for metastatic signet ring cell carcinoma. **e** Myeloma with large eosinophilic immunoglobulin crystals that resemble Auer rods. **f** Myeloma cells with Gaucher cell-like appearance. Myeloma cells may rarely contain numerous, small cytoplasmic crystals that impart a "crumpled tissue paper" like quality to the cytoplasm, resembling the storage histiocytes of Gaucher disease

Fig. 3.8 Dutcher bodies in plasma cell myeloma. Dutcher bodies are not infrequently seen in myeloma and are occasionally very prominent (*arrows*), as in this case (**a** bone marrow biopsy; **b** bone marrow aspirate)

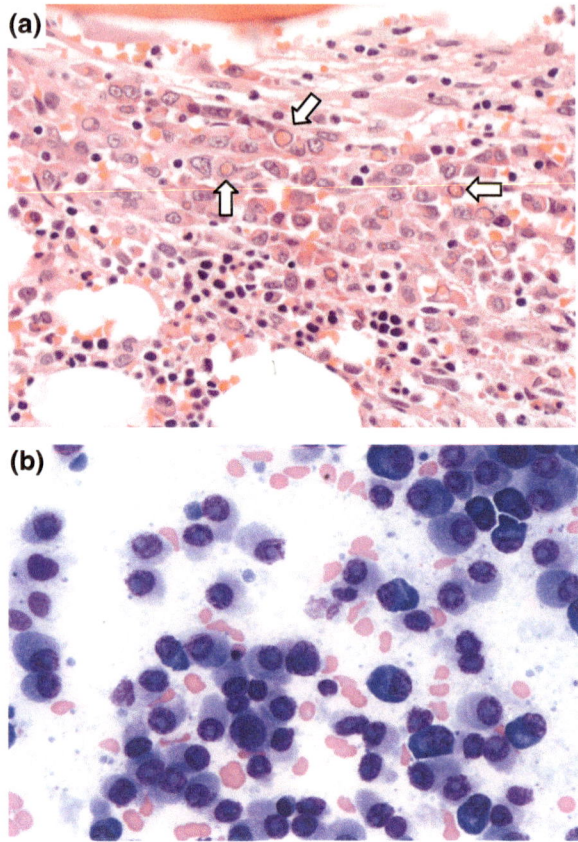

So-called "flame cells" may be prominent in some PCMs (Fig. 3.9) [19]. These cells are so named because of the distinct pink hue of the peripheral cytoplasm of the myeloma cells. This peculiar cytologic feature is associated with synthesis of an IgA paraprotein and is thought to result from the relatively high carbohydrate content of this immunoglobulin isotype [20].

Reactive PCs and most neoplastic PCs have relatively abundant cytoplasm and thus a low nuclear-to-cytoplasmic (N:C) ratio. A subset of myelomas is comprised predominantly of PCs with high N:C ratios, including small lymphocyte-like myeloma by definition (see below for further discussion). In a subset of high-grade myelomas, the tumor cells have relatively scant cytoplasm (Fig. 3.5a), which can obscure their plasmacytic differentiation and make their distinction from other hematologic malignancies, especially large cell lymphoma and acute myeloid leukemia, difficult on a morphologic basis alone.

Neoplastic PCs may rarely manifest hemophagocytic activity (Fig. 3.10) [21–23]. Based on reported cases, there is no known association of hemophago-cytosis with any known genetic lesion in myeloma. Whether such phagocytosis is causally linked to the cytopenias reported in some patients with hemophagocytic

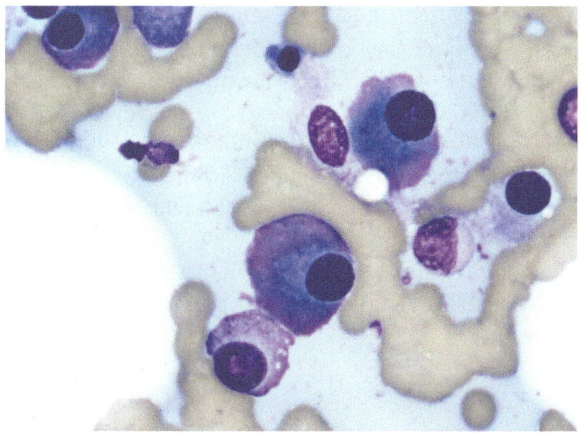

Fig. 3.9 Plasma cell myeloma with flame cells. Several neoplastic PCs containing fuchsia-colored cytoplasm with accentuation in the periphery of the cells

PCM is uncertain. Secondary hemophagocytic lymphohistiocytosis, in which histiocytes rather than the neoplastic PCs are hemophagocytic, occurs only rarely in association with myeloma [24].

Finally, the pattern of bone marrow involvement can provide helpful diagnostic clues that complement the cytologic findings described above. In MGUS and in a

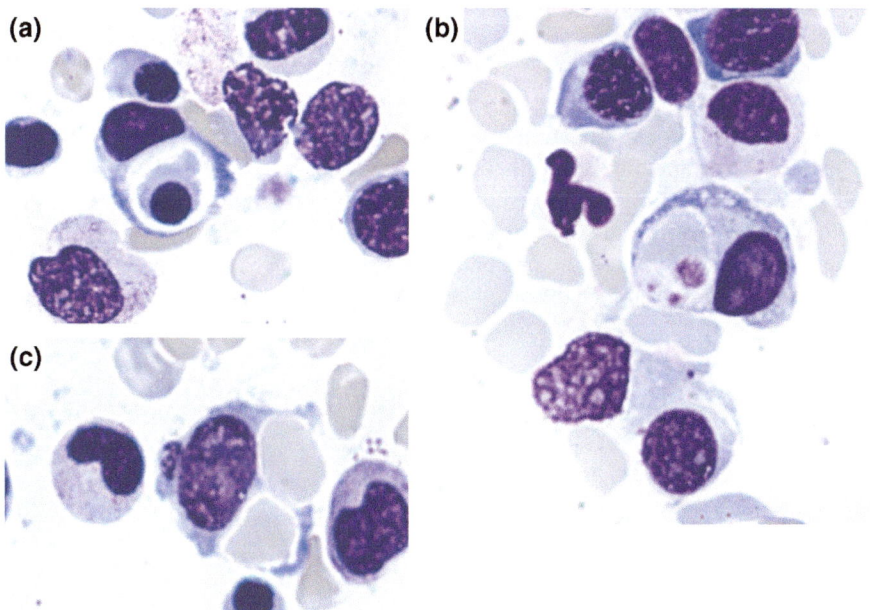

Fig. 3.10 Hemophagocytic plasma cells are rarely observed in myeloma. PCs contain phagocytosed erythroid progenitors (**a**), RBCs and platelets (**b**), and enucleated RBCs (**c**)

subset of myelomas, the clonal PCs are present singly or in small clusters, an interstitial pattern indistinguishable from that seen in reactive plasmacytoses. In most myelomas, however, the degree of bone marrow involvement is much greater, readily permitting distinction from a reactive plasmacytosis. The distribution of myeloma cells is quite variable and includes interstitial, nodular, focal, and diffuse patterns of marrow involvement. Depending on the extent of involvement, nearly all myelomas show at least partial and often subtotal or complete obliteration of the bone marrow architecture. As discussed above, the marked heterogeneity of marrow involvement characteristically seen in myeloma highlights the importance for ample, high-quality marrow sampling at both initial diagnosis and for post-therapy disease monitoring.

Grading of PCM

As with other hematologic malignancies, several grading schemes have been developed for PCM in an attempt to identify those tumors more likely to manifest aggressive clinical behavior. Grading in some systems hinges on the identification and enumeration of minor populations of immature plasmacytic forms (i.e., plasmablasts) in the bone marrow aspirate [25, 26]. In others, grading relies on the bone marrow biopsy findings [27, 28]. For example, the Bartl system is based on the cytomorphology of the neoplastic cells in the bone marrow biopsy and recognizes three grades of myeloma. We have found anecdotally that Bartl grade III (high-grade) myelomas frequently have complex karyotypes and manifest aggressive clinical behavior.

However, the current value of the pathologic grading of myelomas is uncertain. All of the published classification schemata were developed prior to the modern era of myeloma therapeutics (e.g., thalidomide, autologous stem cell transplantation and proteasome inhibitors). Whether such grading retains any independent prognostic value is entirely unknown at present. Nevertheless, it is prudent to appropriately document in the hematopathology report the presence of high-grade cytologic features (e.g., anaplasia, plasmablastic morphology, high proliferative fraction) to alert the treating oncologist to a myeloma that may manifest aggressive clinical behavior.

Unusual PCM Variants

Small Lymphocyte-Like PCM

The small lymphocyte-like morphologic variant (SL-L PCM) comprises approximately 10% of all PCMs and is strongly associated with the t(11; 14) translocation, which results in dysregulated *CCND1* gene and cyclin D1 protein expression [29, 30]. Morphologically, SL-L PCMs are defined by a predominance of neoplastic PCs with relatively scant cytoplasm, a cytologic feature that may result in their misinterpretation as small lymphocytes (Fig. 3.11). In contrast to typical PCM, SL-L PCMs frequently coexpress CD20 and may even express other B-lineage markers such as PAX5 and surface light chain [31]. Thus, SL-L PCM manifests a

Fig. 3.11 Small lymphocyte-like plasma cell myeloma is characterized by neoplastic cells with relatively scant cytoplasm that closely resemble normal mature lymphocytes. While some cases are comprised predominantly of such cells (**a**), others have variable components of Marshalko-type cells together with lymphoplasmacytoid forms (**b**). Appearance of small lymphocyte-like PCM in the bone marrow biopsy (**c**). Small lymphocyte-like morphology is strongly associated with the t(11;14)

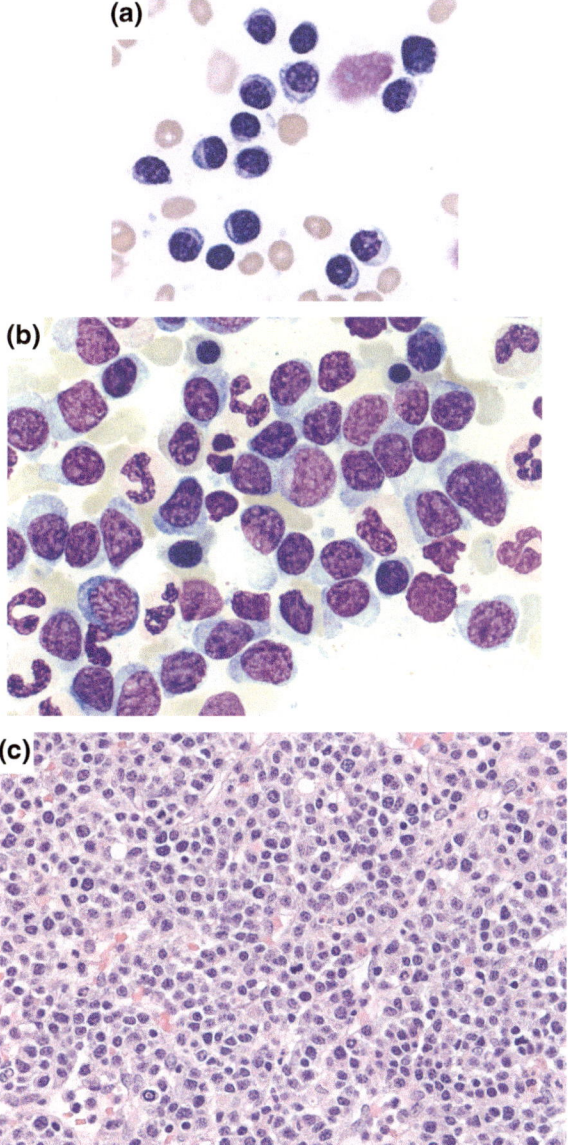

constellation of morphologic and immunophenotypic findings transitional between that of mature B-cell lymphoma and PCM, potentially leading to its misdiagnosis as chronic lymphocytic leukemia/small lymphocytic lymphoma or other low-grade B-cell lymphomas with plasmacytic differentiation, such as lymphoplasmacytic lymphoma (LPL) or marginal zone lymphoma (MZL) [32]. When confronted with a SL-L PCM, cyclin D1 immunohistochemistry, t(11;14) FISH and correlation with serum protein analyses can be critical in discriminating between these diagnostic possibilities. SL-L PCM is usually cyclin D1 positive due to the high incidence of the t(11;14) and is usually accompanied by IgG or IgA paraproteinemia. By contrast, LPL and MZL are uniformly cyclin D1 negative and are usually associated with an IgM paraprotein [33].

Nonsecretory PCM

Although a monoclonal immunoglobulin paraprotein is a hallmark of PCM, a subset of myelomas manifests defective immunoglobulin synthesis or secretion. Most of these tumors retain some degree of immunoglobulin secretion and are thus designated as hyposecretory. Rare PCMs show a complete block in paraprotein biosynthesis as assessed by any currently available serum protein analytic technique and thus fulfill the strictest criteria for true nonsecretory myeloma (NS-PCM). Historically, the distinction between hyposecretory and bona fide NS-PCM has been imprecisely defined, given the availability of relatively low-sensitivity assays for serum or urine paraprotein detection. Serum protein electrophoresis (SPE) has a sensitivity of 500–2000 mg/L, whereas the sensitivity of immunofixation electrophoresis is at most tenfold greater, depending on the particular paraprotein [34]. Newer serum free light chain analytic techniques are much more sensitive, affording the detection of serum free light chains in the low mg/dl range [35]. Quite predictably the incidence of stringently defined NS-PCM has dropped significantly since the introduction of more sensitive protein analytic methodology.

With a constellation of clinical findings that includes lytic bone lesions, cytopenias and no detectable paraprotein, NS-PCMs are often clinically misdiagnosed as metastatic malignancies, particularly metastatic prostate or breast carcinoma depending on the patient's gender. Fortunately from a pathologic perspective, NS-PCMs are far less challenging diagnostically as they manifest typical histopathologic features of PCM (Fig. 3.12). Interestingly, the majority of NS-PCMs contains the t(11;14) and as such express cyclin D1 [36, 37]. NS-PCM frequently shows no detectable immunoglobulin light chain expression, as assessed by either immunohistochemistry or in situ hybridization (Fig. 3.12d, e) [37]. This suggests that the nonsecretory state of these myelomas is due to either impaired light chain gene transcription or expression of a truncated or destabilized light chain RNA transcript rather than some downstream defect in immunoglobulin protein secretion; however, the molecular basis for this is not well understood at present [38, 39].

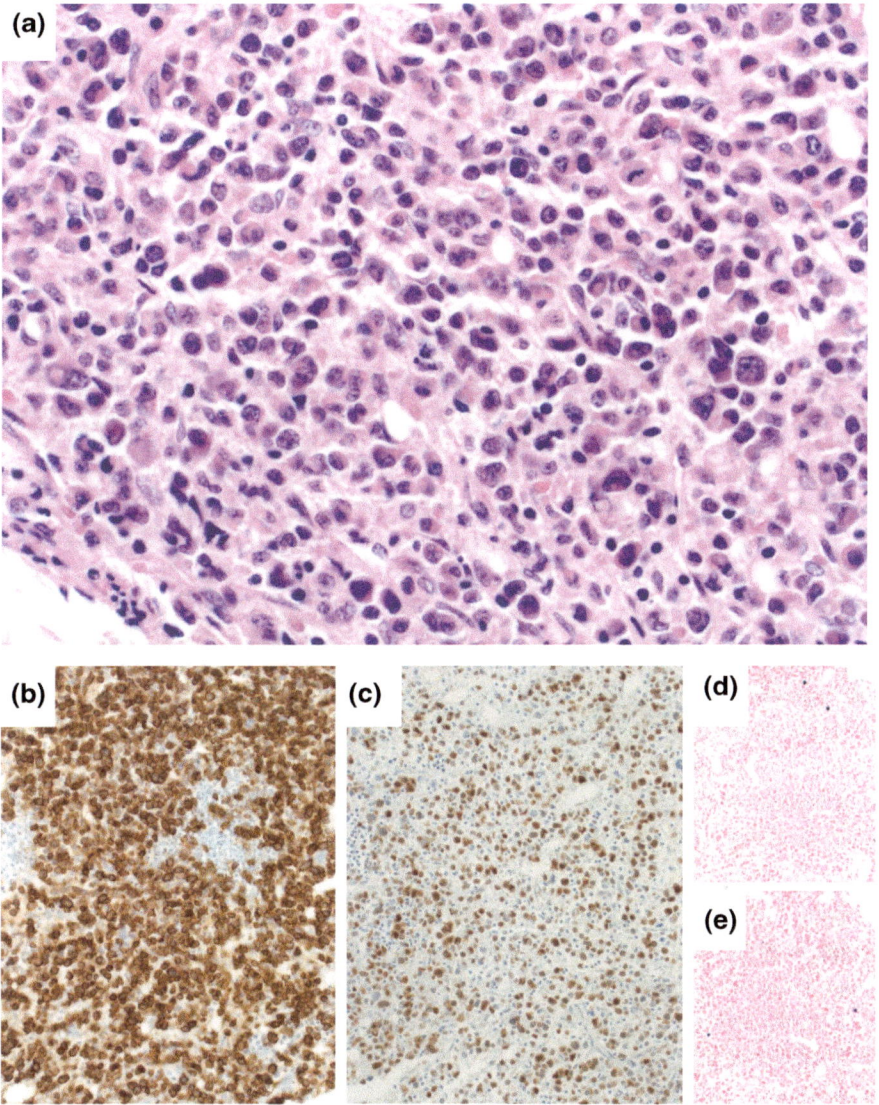

Fig. 3.12 Nonsecretory plasma cell myeloma. Diffuse infiltrate of pleomorphic neoplastic PCs (**a**) that are positive for CD138 (**b**) and cyclin D1 (**c**). In situ hybridization for kappa (**d**) and lambda (**e**) confirms absence of light chain expression by the neoplastic PCs; rare reactive light chain expressing PCs are present. FISH confirmed the presence of the t(11;14) (not shown)

Plasma Cell Leukemia (PCL)

PCL is a recognized variant form of PCM in the WHO classification. Primary PCL (i.e., leukemic disease at initial presentation) is rare, accounting for approximately 5% of all PCM patients, and is associated with a very poor clinical outcome [40, 41]. Primary PCL is often associated with extramedullary disease [41, 42]. It is less frequently associated with lytic bone lesions in some but not all studies [42, 43]. Synthesis of light chain only is detected in 20–44% of primary PCL, an incidence much higher than the 15% light chain paraproteinemia occurring in typical PCM [41, 44]. Secondary PCL is more common and typically develops in patients in late clinical stage with refractory, disseminated disease.

PCL is usually diagnostically straightforward, as most cases develop in patients with known PCM. The circulating tumor cells usually manifest some degree of overt plasmacytic differentiation (Fig. 3.13a), although some cases show marked cytologic heterogeneity (Figs. 3.14 and 3.15). The neoplastic cells in some PCLs may possess scant cytoplasm, more closely resembling small lymphocytes and thus be misinterpreted as either a reactive lymphocytosis or as a lymphoid leukemia such as chronic lymphocytic leukemia (Fig. 3.13b). In some poorly differentiated PCLs, the leukemic cells may closely resemble myeloid blasts or peripheralized large cell lymphoma tumor cells (Fig. 3.13c). Such cases are especially challenging diagnostically, particularly in the absence of any other laboratory or radiologic information.

PCL shares many immunophenotypic similarities with PCM, although there are a few notable differences. In comparison with PCM, PCL is more likely to be negative for CD56 and CD117 and more commonly express CD20 [42]. The latter finding reflects the high incidence of the t(11;14) in primary PCL [43], a genetic lesion strongly associated with aberrant CD20 expression. The relatively infrequent expression of the adhesion molecule CD56 in PCL is believed to account, at least in part, for the reduced adhesion that contributes to the peripheralization of neoplastic cells in PCL. The diagnosis of PCL is usually confirmed by flow cytometry; appropriate detection of the neoplastic PCs requires at a minimum inclusion of CD38 in the flow cytometry antibody panel, and preferably also CD138 and cytoplasmic light chain analysis (see Chap. 5).

Primary Amyloidosis and PCM with Associated Amyloidosis

The extracellular deposition of Ig light chains in the form of insoluble molecules with a β-pleated sheet configuration is the defining pathogenetic event in primary amyloidosis [45]. End-organ damage resulting from this deposition, which can occur at virtually any anatomic site (most commonly kidney, heart, GI tract and peripheral nerves), accounts for the protean clinical manifestations of amyloidosis.

Fig. 3.13 Plasma cell leukemia. **a** Typical leukemic PC with overt plasmacytic differentiation. **b** PCL comprised of small lymphoid cells with subtle plasmacytic features. These cases may be misdiagnosed as a small B-cell leukemia, such as chronic lymphocytic leukemia. **c** PCL comprised of large cells with cytologic features resembling those of myeloid blasts or peripheralized large cell lymphoma. Figure kindly provided by Dr. Marwan Yared

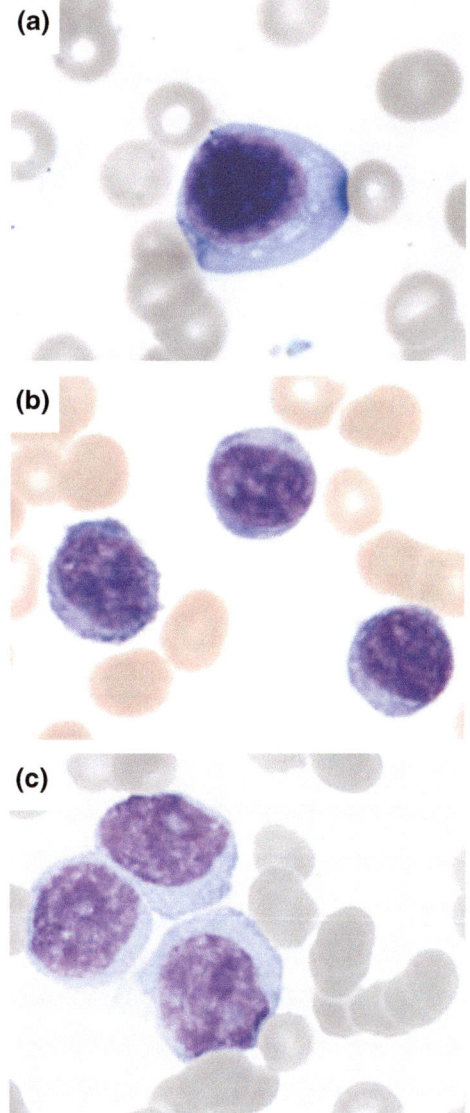

In the bone marrow, primary amyloidosis is characterized by amyloid deposits accompanied frequently but not invariably by a plasmacytosis. Amyloid is deposited either within/around blood vessels or within the marrow interstitium (Fig. 3.16a, b). The deposition of amyloid can be sufficiently extensive as to form space-occupying lesions, which may elicit a foreign-body giant cell reaction. Many laboratories routinely perform a periodic acid Schiff (PAS) stain on bone marrow biopsies. An advantage of the PAS stain is that it highlights subtle amyloid deposits that may be undetectable in standard H&E sections (Fig. 3.16c).

Fig. 3.14 Plasma cell
leukemia. This image from a
single high power field
highlights the marked
cytologic heterogeneity of the
circulating neoplastic cells in
some cases of plasma cell
leukemia. Neoplastic cells
with obvious plasmacytic
features (*open arrows*) as well
as cells resembling myeloid
blasts (*filled arrow*) and
circulating large cell
lymphoma cells
(*cross-hatched arrow*).
Figure kindly provided by Dr.
Marwan Yared

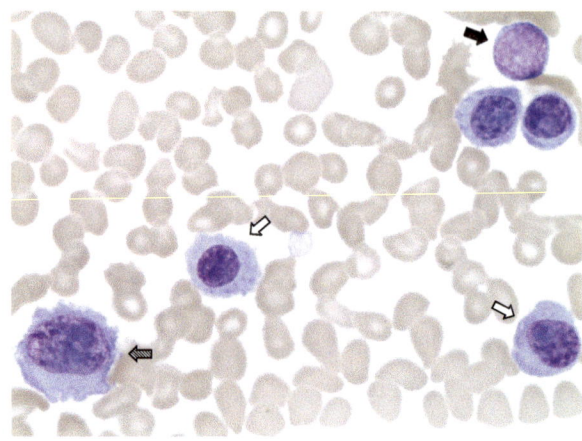

Fig. 3.15 Plasma cell
leukemia. **a** In some cases of
PCL, neoplastic cells
concentrate along the
feathered edge and may be
significantly underrepresented
in the monolayer portion of
the blood film. Failure to
recognize the neoplastic
PCs may result from
neglecting to examine the
feathered edge in such cases.
b Neoplastic PCs are apparent
at higher magnification

(a)

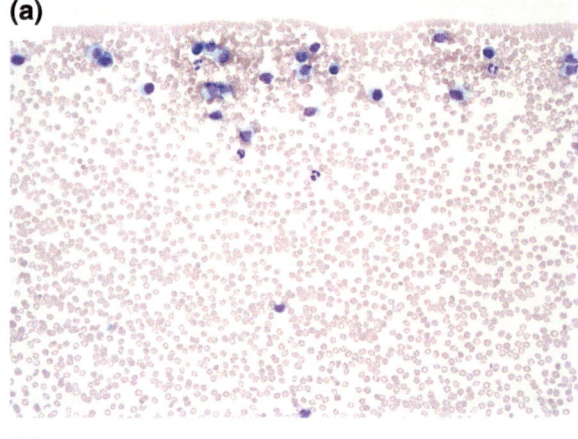

(b)

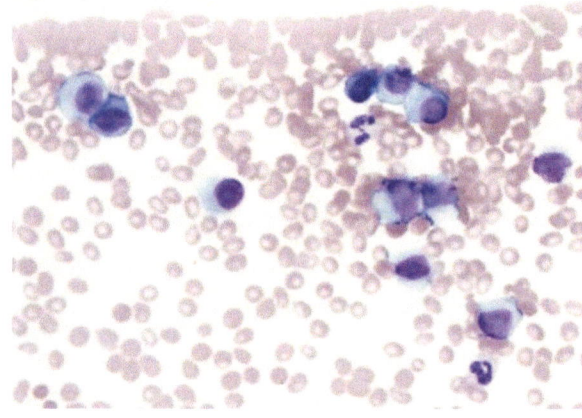

Fig. 3.16 Plasma cell myeloma accompanied by primary amyloidosis. **a** Prominent interstitial amyloid deposits are evident under low-power examination. **b** These deposits have an amorphous, lightly eosinophilic appearance in H&E sections, typical of amyloid. A prominent infiltrate of neoplastic myeloma cells is present in the intervening marrow. **c** In this case, there is less prominent amyloid deposition, most of which has an intramural/perivascular distribution and is highlighted in this PAS stain

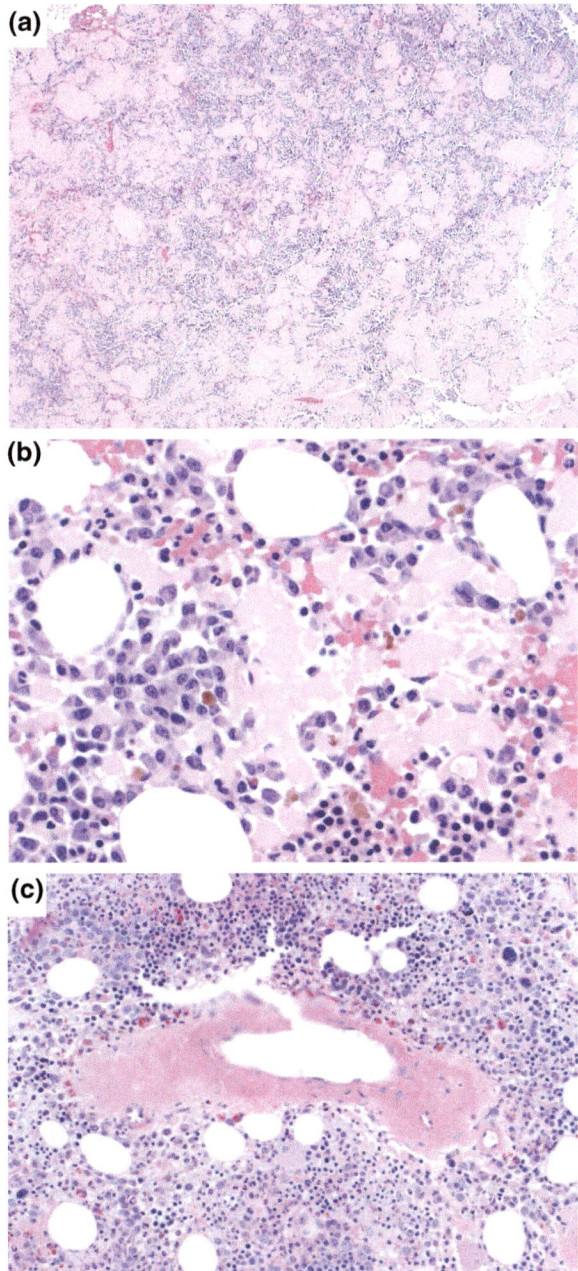

Amyloid deposits are usually inconspicuous in the bone marrow aspirate, but they may be more obvious in the setting of marked amyloid deposition. In such cases, the amyloid often imparts an almost geometric appearance to the bone marrow spicules that is readily apparent under low-power examination (Fig. 3.17a). In Wright-Giemsa stained aspirate smears, amyloid has a basophilic appearance (Fig. 3.17b).

The Congo red stain highlights amyloid deposits, which manifest the characteristic "apple green" birefringence when examined under polarized light (Fig. 3.18a, b). A Congo red stain should be performed any time there is clinical suspicion of amyloidosis irrespective of whether amyloid deposits are evident in routine H&E sections. In addition to a confirmatory Congo red stain, amyloid subtyping should be performed if sufficient tissue is available. While immunohistochemistry has historically been used to determine amyloid subtype, mass

Fig. 3.17 Appearance of amyloid in the bone marrow aspirate. **a** Low-level deposition of amyloid is usually inconspicuous in routine Wright-Giemsa stained aspirate smears. Note the geometric appearance of the spicules. **b** At higher power, amyloid has a lightly basophilic, amorphous appearance (*arrows*)

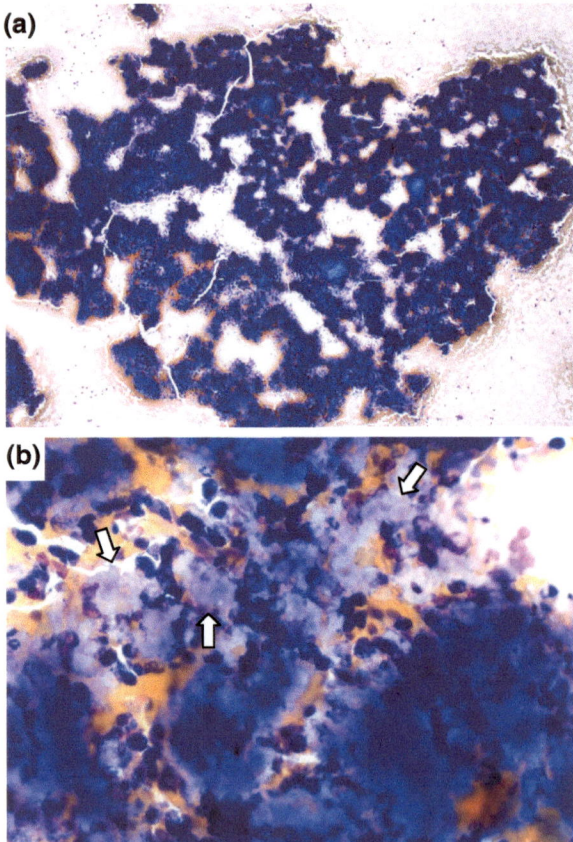

Fig. 3.18 Plasma cell myeloma accompanied by primary amyloidosis. **a** This perivascular amyloid deposit is strongly congophilic. **b** Amyloid deposits manifest so-called "apple green" birefringence when examined under polarized light

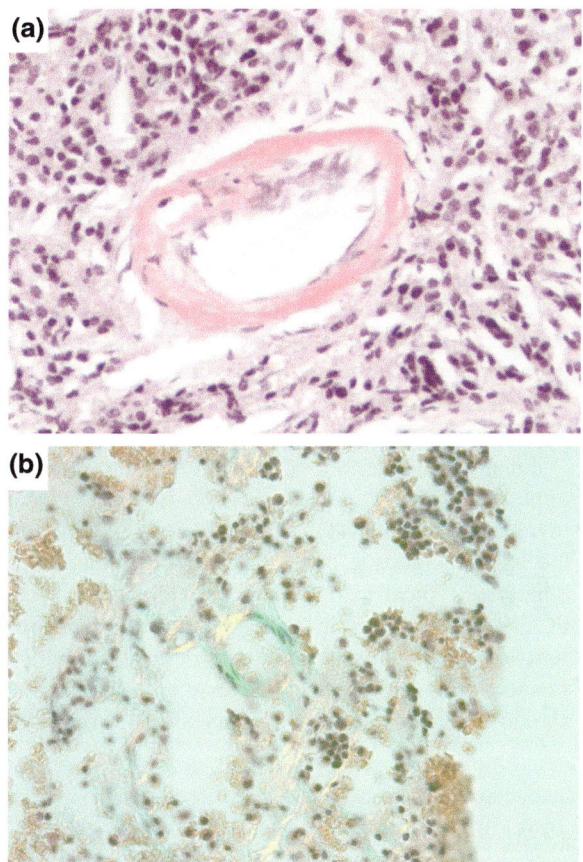

spectrometry is now considered the gold standard methodology for amyloid analysis [46]. Mass spectrometry not only confirms the presence of amyloid but also permits unequivocal identification of the amyloidogenic protein. Even in the setting of an established PC dyscrasia (MGUS or PCM), mass spectrometry should be undertaken to confirm the amyloid subtype, since MGUS is relatively common in the general population and may coexist with other types of non-light chain amyloidosis, for which the therapy is quite different than that for primary amyloidosis.

An important consideration to keep in mind during the diagnostic workup of a case of primary amyloidosis is that the amyloidogenic PC clone often comprises only a subset of the total PC population. Thus, the presence of "contaminating" resident bone marrow PCs can make confirmation of PC clonality by light chain immunohistochemistry or in situ hybridization difficult if not impossible. In such cases, phenotypic characterization of the amyloidogenic PC clone, including confirmation of its light chain restriction, is readily achieved by flow cytometry.

Most cases of primary amyloidosis are characterized by only a modest bone marrow plasmacytosis, usually less than 10% of the cellularity; however, more prominent plasmacytosis raises the possibility of concurrent amyloidosis and PCM. PCM and primary amyloidosis coexist in a subset of patients with otherwise typical myeloma. Interestingly, amyloid deposits can be detected in up to 40% of patients with PCM despite the fact that only 10–15% have clinically overt amyloidosis [47, 48]. This suggests that low-level amyloid deposition may be clinically insignificant in some myeloma patients. It is important to note that primary amyloidosis may develop in association with other lymphoid malignancies, including lymphoplasmacytic lymphoma and MALT lymphoma, albeit more often at extramedullary sites.

Finally, one important caveat of the current WHO classification is that no minimum level of monotypic PCs is specified for a diagnosis of symptomatic myeloma. This reflects the fact that very low levels of neoplastic PCs are present in the initial diagnostic bone marrow specimen in a small minority of patients with otherwise typical symptomatic myeloma. The frequency of clonal bone marrow PCs is similarly low in primary amyloidosis. However, patients with organ dysfunction due solely to amyloid deposition with no other evidence of typical myelomatous end-organ damage (e.g., anemia, osteolytic lesions, hypercalcemia) should be diagnosed with primary amyloidosis rather than symptomatic myeloma.

PCM with Associated Crystal Storing Histiocytosis

PCM is rarely accompanied by a reactive histiocytic hyperplasia in which the histiocytes containing prominent crystalline immunoglobulin inclusions or so-called crystal storing histiocytosis (CSH) [49]. In such cases, the neoplastic PCs usually manifest kappa light chain restriction and contain abundant immunoglobulin inclusions, which often have a needle-like appearance resembling Auer rods (Fig. 3.19a). Crystal formation is thought to reflect the abnormal protein folding of a mutant immunoglobulin molecule [50]. While PC light chain restriction is readily demonstrated by IHC, the immunoglobulin crystals proper may not be reactive, presumably because of epitope masking due to their crystalline nature.

In some cases, the CSH can be sufficiently florid as to mimic a storage disorder, such as Gaucher disease, masking the underlying PC neoplasm (Fig. 3.19b). However, close examination reveals that the histiocytes in CSH are distended with well-delineated crystals (Fig. 3.19c) and lack the classic striated-appearing cytoplasm characteristic of Gaucher disease. Of note, CSH may occur, albeit less

Fig. 3.19 Plasma cell myeloma and concurrent crystal storing histiocytosis **a** Several PCs containing immunoglobulin crystals are present. **b** Prominent histiocytic infiltrate is present in the bone marrow biopsy, mimicking Gaucher disease at low magnification. **c** The crystalline nature of the material within the histiocyte cytoplasm is apparent at high magnification. In this particular case, the histiocytic proliferation overshadows the admixed neoplastic myeloma cells (arrows)

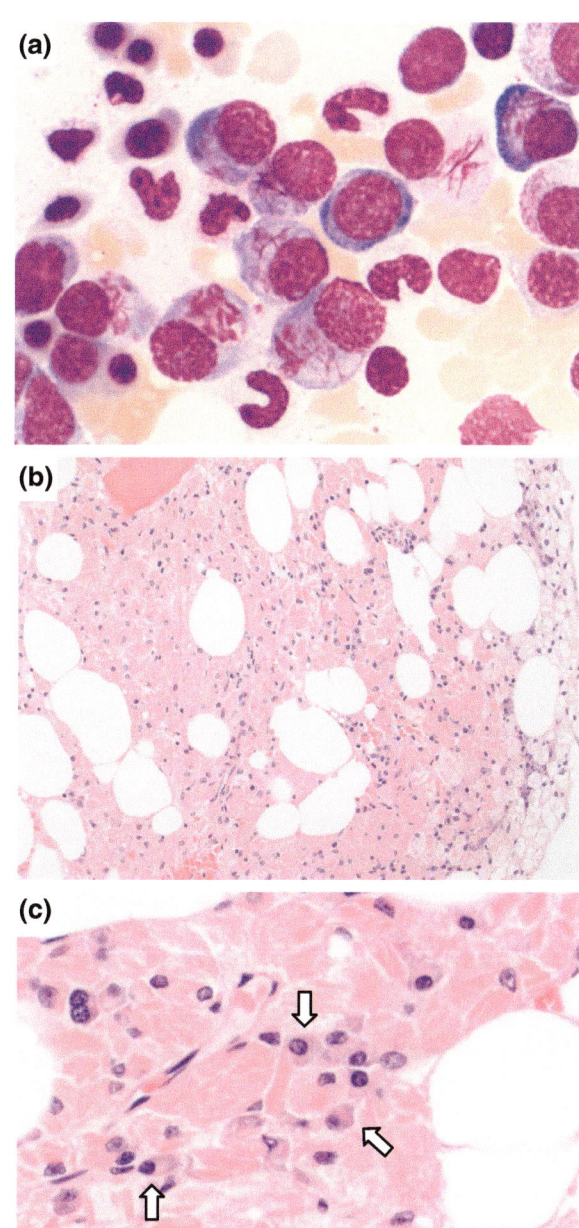

Fig. 3.20 Plasma cell myeloma with myelofibrosis. **a** Diffuse infiltrate of myeloma cells is accompanied by overt fibrosis, most evident in the upper right hand corner. **b** Reticulin staining confirms the marked, diffuse increase in reticulin fiber deposition

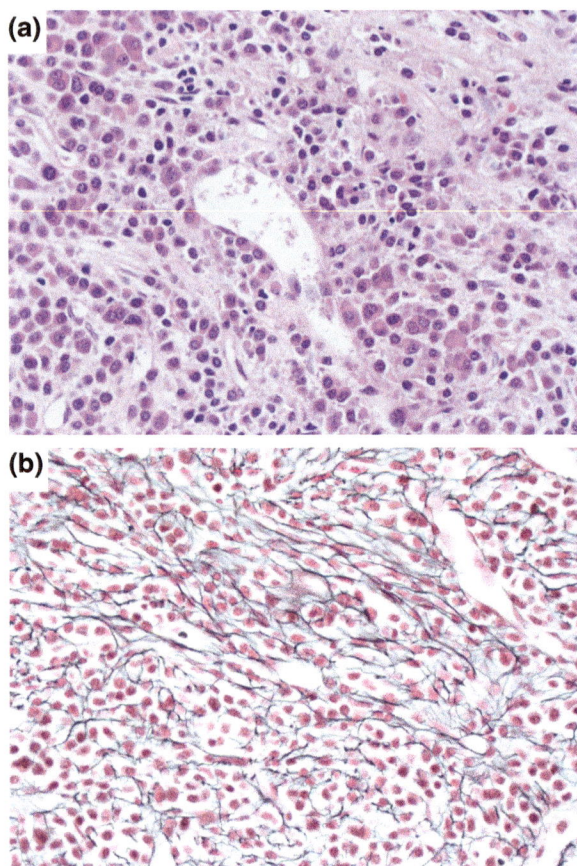

frequently, in association with lymphomas with plasmacytic differentiation (e.g., lymphoplasmacytic lymphoma and extranodal marginal zone lymphoma), reactive disorders and even non-hematolymphoid neoplasms such as inflammatory myofibroblastic tumors [49, 51].

PCM with Myelofibrosis

PCM is frequently accompanied by patchy, low-grade reticulin fibrosis; however, PCM is occasionally associated with overt myelofibrosis (Fig. 3.20) [52–54]. In such instances, it may be difficult or impossible to obtain a satisfactory bone marrow aspirate, resulting in significant underrepresentation of the neoplastic myeloma cells in the aspirate smears. Due to the cytologic atypia, fibrosis, and

Fig. 3.21 Plasma cell myeloma with fibrosis. **a** In this post-therapy bone marrow biopsy, a deposit of residual myeloma with associated fibrosis was focally present. **b** The fibrosis largely obscures the plasmacytic differentiation of the residual myeloma cells. **c** Kappa light chain in situ hybridization confirms the plasmacytic lineage and clonality of the cells within the fibrotic focus and highlight the paucity of neoplastic PCs in the remainder of the biopsy

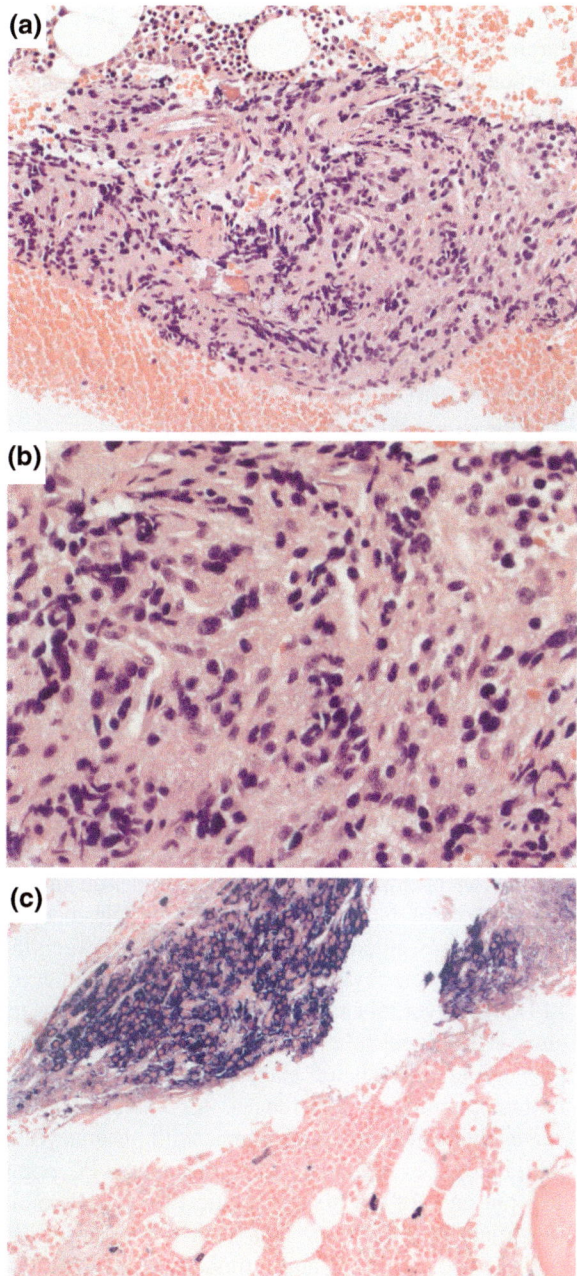

crush artifact that are often present in the biopsy, such cases can mimic metastatic carcinoma. Myelofibrosis may render myeloma cells more refractory to chemotherapy, consistent with the emerging concept that microenvironment perturbation is an important pathogenetic factor in PCM [53, 54]. Indeed, residual myelomatous deposits following chemotherapy are sometimes associated with significant fibrosis (Fig. 3.21). However, the prognostic impact, if any, of myelofibrosis in PCM has not been rigorously evaluated since the advent of newer myeloma therapeutics such as thalidomide and proteasome inhibitors.

Association of Plasma Cell Myeloma with Other Malignancies

PCM or MGUS may occasionally coexist with other malignancies (Box 3.4). Of the hematologic malignancies, chronic lymphocytic leukemia (CLL) and its precursor, monoclonal B-cell lymphocytosis (MBL), are most commonly seen together with PCM/MGUS (Fig. 3.22). Fortunately, the development of a PC dyscrasia and a B-cell lymphoproliferative disorder (LPD) is metachronous in most patients and as such is relatively straightforward from a diagnostic perspective. Molecular analysis has shown that concurrent PCM and CLL represents two distinct diseases and are clonally unrelated in most cases [55–57]. Simultaneous detection of clonal PC *and* B-cell populations poses a greater diagnostic challenge in a patient with no known history of either a PC dyscrasia or B-cell LPD. In this setting, the diagnostic possibilities include two coexisting neoplasms (e.g., CLL/MBL and a PC dyscrasia) versus a single neoplasm, namely a B-cell lymphoma with plasmacytic differentiation (e.g., LPL, MALT lymphoma or splenic marginal zone lymphoma). This distinction is important as it has therapeutic implications. Flow cytometric characterization of both the lymphoid and plasmacytic components, including light chain analysis, often helps to distinguish between these possibilities. Different light chain restriction obviously favors a diagnosis of concurrent PC dyscrasia and B-cell LPD. However, shared light chain restriction by the PCs and B-cells is not conclusive for clonal relatedness and does not exclude coexisting neoplasms. Detailed phenotypic analysis of each population is often helpful. There are significant immunophenotypic differences, for example, between the PCs of PCM/MGUS versus those of B-cell lymphomas with plasmacytic differentiation [58–60]. Careful morphologic evaluation, particularly of the PC population, may be helpful as well. The presence of overtly dysplastic PCs (e.g., prominent nucleoli) is uncommon in low-grade B-cell lymphomas with plasmacytic differentiation and favors a diagnosis of coexisting PC and lymphoid neoplasms. For challenging cases, correlation with radiologic, cytogenetic/FISH and serum protein studies may be necessary for definitive diagnosis.

Fig. 3.22 Concurrent bone marrow involvement by plasma cell myeloma and chronic lymphocytic leukemia. **a** Low-power reveals a diffuse infiltrate of myeloma cells with focal aggregate of small lymphocytes (*dashed line*). The myeloma cells express CD138 (**b**) with kappa light chain restriction (**c**). The CLL cells are positive for CD20 (**d**) and CD5 (**e**)

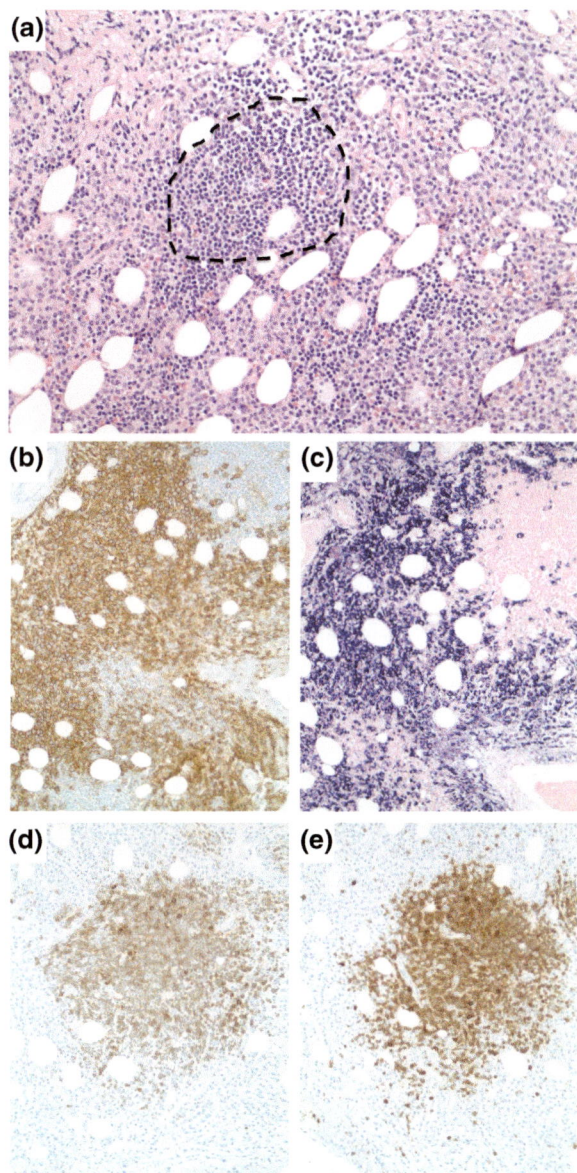

Box 3.4: Most common synchronous/metachronous neoplasms coexisting with PCM

Small B-cell neoplasms

Monoclonal B-cell lymphocytosis
Chronic lymphocytic leukemia/small lymphocytic lymphoma
Marginal zone lymphoma
Miscellaneous (follicular lymphoma, mantle cell lymphoma)

Therapy-related myeloid malignancies

Therapy-related AML
Therapy-related MDS

Metastatic carcinoma (prostate, breast, lung)
Mast cell disease

PCM rarely coexists with other non-lymphoid malignancies. Concurrent myeloma and metastatic carcinoma are occasionally encountered (Fig. 3.23). CD138 is expressed by PCM and many carcinomas; thus, CD138 IHC should always be interpreted with caution, particularly in cases of poorly differentiated PCM. In difficult cases, cytokeratin, MUM1, and light chain IHC is of utility in confirming the presence of concurrent PCM and metastatic carcinoma. Rarely, systemic mastocytosis may occur in association with PCM (Fig. 3.24) [61–63]. In some instances, the mast cell infiltrate may be sufficiently prominent to mask the PC neoplasm.

Finally, myeloma patients typically have extensive exposure to cytotoxic chemotherapy, including alkylating agents such as melphalan, and are consequently at risk for the development of therapy-related myelodysplasia and AML (Fig. 3.25). In addition to assessment for residual PCM, post-therapy marrows should be carefully scrutinized for the presence of dysplasia or increased blasts, particularly in those patients who have cytopenias unexplained by persistent PCM or recent chemotherapy. Metaphase cytogenetics or FISH to detect common structural or numeric abnormalities (trisomy 8, monosomy 5/deletion 5q, monosomy 7/deletion 7q, deletion 20q) may be helpful in confirming a diagnosis of therapy-related myeloid malignancy. Because therapy-induced dyspoietic changes can be difficult to distinguish from bona fide low-grade myelodysplasia, more extensive genomic analysis such as targeted gene exon sequencing may be informative in cases lacking the common MDS-related cytogenetic lesions above.

Fig. 3.23 Concurrent bone marrow involvement by plasma cell myeloma and metastatic prostate adenocarcinoma. **a** A large deposit of metastatic carcinoma is present in the upper portion of the biopsy. **b** CD138 immunohistochemistry labels both the carcinoma and the myeloma cells; however, the latter are much more intensely CD138 positive

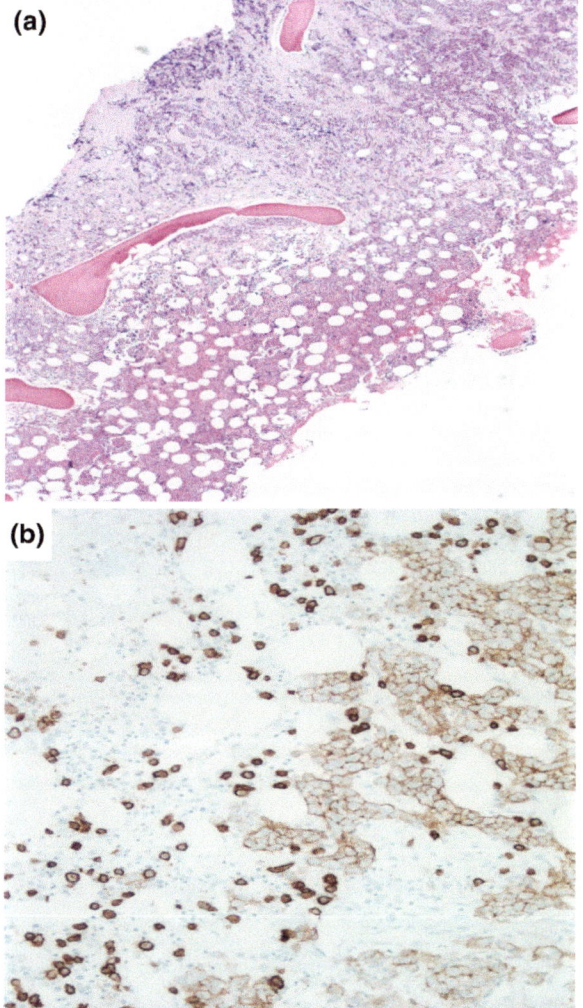

Immunohistochemical Evaluation of Plasma Cell Neoplasms

Immunophenotypic characterization plays an important role in the initial diagnosis of PCM and in the evaluation of follow-up bone marrow and other specimens in patients with an established diagnosis of myeloma. Immunohistochemistry (IHC) has several important diagnostic applications in this setting. First, IHC is helpful in confirming the plasmacytic lineage of the neoplastic cells, particularly when confronted with a poorly differentiated plasma cell neoplasm. Second, IHC frequently permits a more accurate quantitation of the neoplastic PC infiltrate than

Fig. 3.24 Concurrent bone
marrow involvement by
plasma cell myeloma and
systemic mast cell disease.
a Multiple aggregates of mast
cells are present in the bone
marrow biopsy, one of which
is shown in the left-hand
portion of the panel. Small
clusters of neoplastic PCs are
present at the periphery of the
mast cell aggregate and in
intervening marrow (*arrows*).
b The mast cells are strongly
positive for CD117 and
tryptase (not shown).
c CD138 highlights myeloma
cells surrounding the mast cell
aggregate and in the
intervening marrow, which
showed monotypic kappa
light chain expression (not
shown)

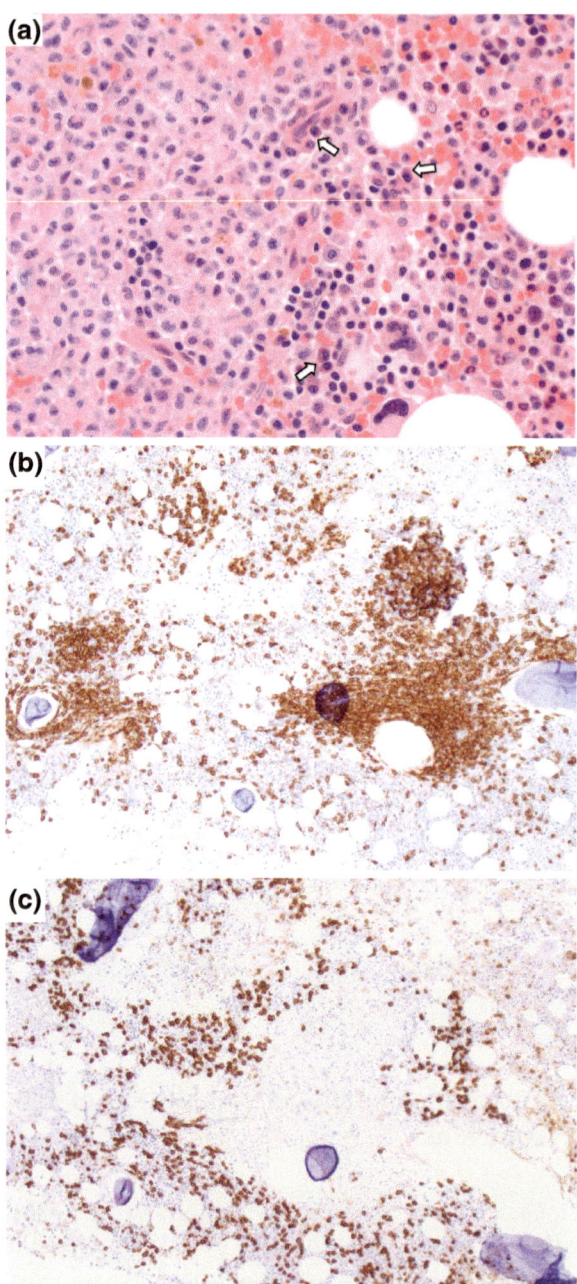

Fig. 3.25 Concurrent plasma cell myeloma and therapy-related myeloid malignancy. A 70-year old man with a 6 year history of myeloma and extensive history of cytotoxic chemotherapy developed unexplained cytopenias. Bone marrow examination revealed persistent PCM (**a**) accompanied by pronounced trilineage dysplasia (**a** and **b**). Erythroid elements show megaloblastoid maturation and prominent multinucleation and nuclear lobation (**a** and **b** *filled arrows*). Dysplastic hypolobated megakaryocytes are present (**b** *open arrow*). **c** Myeloid blasts were increased, accounting for 24% of analyzed cells by flow cytometry, consistent with therapy-related acute myeloid leukemia. Cytogenetics revealed a complex karyotype that included del(5)(q21q34) and inv(3)(q21q26.2), common genetic lesions in therapy-related myeloid neoplasms

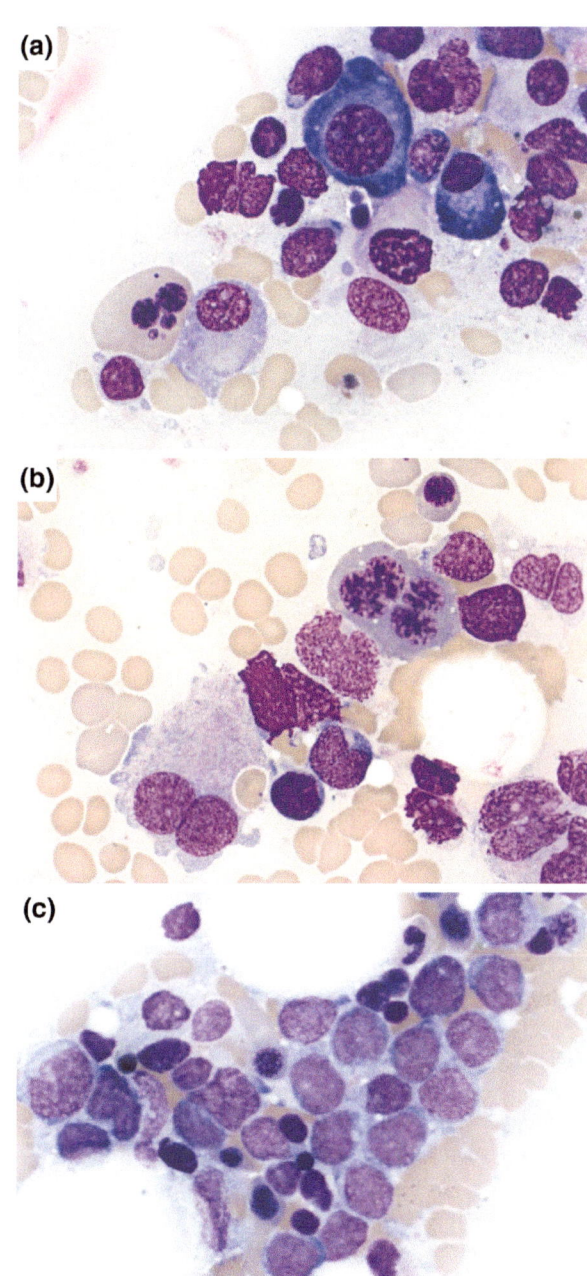

that afforded by evaluation of the H&E-stained biopsy sections or aspirate smears alone. Third, some immunohistochemical markers, most notably cyclin D1, are associated with specific myeloma-related genetic lesions and thus have particular diagnostic value in distinguishing myeloma from its mimics, low-grade B-cell lymphomas with plasmacytic differentiation.

Similar to their normal counterparts, neoplastic PCs express several cell surface and cytoplasmic antigens associated with plasmacytic differentiation. The antigens of greatest utility for IHC include CD138, CD38, CLIMP-63 (VS38c), MUM1, and the Ig heavy and light chains (Box 3.5).

Box 3.5: Plasma cell myeloma immunohistochemistry—caveats and practical considerations

- Most PCM are strongly, diffusely CD138 positive
- Although CD138 is a highly sensitive and specific marker for plasmacytic differentiation in hematolymphoid malignancies, a wide array of carcinomas are also CD138 positive
- MUM1 or immunoglobulin immunohistochemistry should be used with poorly differentiated PCM to confirm the diagnosis
- Cyclin D1 is helpful in distinguishing small lymphocyte-like PCM from its mimics
- Hyposecretory PCM may be light chain negative by immunohistochemistry

---

CD138

CD138 is an excellent immunohistochemical marker for PCs. CD138 is highly expressed by virtually all PCs, benign and malignant, and within the hematolymphoid system, it is a highly specific marker of plasmacytic differentiation [64–66]. CD138 IHC permits more accurate quantitation of myeloma cells, making it particularly useful for detection of low levels of residual disease in post-therapy biopsies and in the workup of small lymphocyte-like myelomas where the distinction of lymphoplasmacytoid PCs from reactive lymphocytes and erythroid cells in routine H&E sections can be challenging [67, 68].

A few important caveats should be kept in mind, however, when interpreting CD138 IHC stains. First, while CD138 expression is indicative of plasmacytic differentiation, it is not diagnostic of PCM. CD138 expression is present in a wide array of lymphomas that manifest varying degrees of plasmacytic differentiation, including lymphoplasmacytic lymphoma, extranodal marginal zone lymphoma, plasmablastic lymphoma, and rare ALK-positive diffuse large B-cell lymphoma (DLBCL) variants [1]. The extent of CD138 expression is variable in these

Fig. 3.26 Extracellular deposition of CD138 (syndecan). CD138 immunohistochemistry highlights PCs (*lower right quandrant*) as well as abundant extracellular CD138 not associated with intact cells (most of immunoreactivity in *left side* of image)

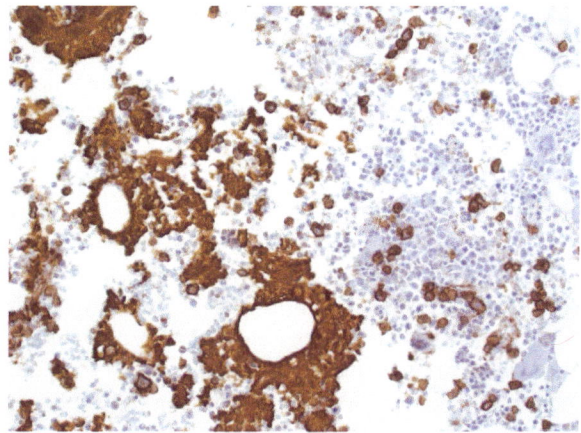

lymphomas, reflecting the extent of plasmacytic differentiation. Second, CD138 expression is occasionally observed in lymphomas otherwise lacking plasmacytic differentiation, including classical Hodgkin lymphoma and anaplastic large cell lymphoma [65, 66]. In such instances, the CD138 positivity is typically weak or focal, unlike the strong, uniform CD138 expression typical of PCM. Third, CD138 is expressed by most normal epithelia irrespective of anatomic site and approximately one-half of all pulmonary, breast, gastrointestinal, and genitourinary carcinomas [69, 70]. Thus, CD138 positivity alone is insufficient to unequivocally establish the plasmacytic origin of poorly differentiated tumors, particularly, those arising at extramedullary sites. In these instances, plasmacytic differentiation of the tumor cells should be confirmed by IHC for other PC-lineage markers.

Finally, CD138 may undergo heparanase-mediated shedding from the cell surface of myeloma cells, resulting in extracellular deposition where it can be detected by immunohistochemistry (Fig. 3.26) [71, 72]. Such extracellular CD138 deposits may persist after chemotherapy-induced clearance of myeloma cells from the marrow. Thus, CD138 immunostains should always be carefully evaluated to ensure that the observed immunoreactivity is cell-associated and does not simply reflect stromal deposition.

MUM1

The transcription factor MUM1, encoded by the *IRF4* gene, is highly expressed in nearly all PCMs, where it plays a critical pathogenetic role [73]. MUM1 is normally expressed in PCs and a subset of germinal center B-cells [74]. MUM1 is more widely expressed in hematolymphoid neoplasia than CD138, being expressed in classical Hodgkin lymphoma and several subtypes of B-cell and T-cell lymphoma in addition to myeloma and lymphomas with plasmacytic differentiation [74, 75]. In contrast to CD138, MUM1 is not expressed in epithelial malignancies. Thus, the primary utility of MUM1 IHC in the diagnostic workup of myeloma is for confirmation of plasmacytic differentiation and distinction of poorly differentiated myelomas from carcinomas and other tumor types.

CLIMP-63

The antibody VS38c recognizes cytoskeleton linking membrane protein-63 (CLIMP-63; also known as p63) [76, 77]. CLIMP-63 localizes to the rough endoplasmic reticulum where it mediates docking of the RER to microtubules. Prior to identification of the cognate antigen, VS38c was shown to be a useful antibody for the immunohistochemical detection of PCs [78]. Normal PCs and essentially all PCMs, plasmacytomas, and lymphoplasmacytic lymphomas are strongly VS38c positive, whereas other B-cell lymphomas are either negative or only weakly positive. Although the expression of CLIMP-63 in non-hematolymphoid malignancies has not been well characterized, it should be noted that more than 90% of melanomas express CLIMP-63 [79].

Immunoglobulin Heavy and Light Chains

IHC for immunoglobulin heavy chain (HC) and light chain (LC) is frequently helpful in the evaluation of PC neoplasms, serving both to confirm plasmacytic differentiation and clonality. In addition to its diagnostic utility, immunoglobulin IHC confirms and sometimes helps to clarify serum protein analyses, particularly when such studies yield ambiguous results or when the serum paraprotein level is very low. Evaluation of HC expression is diagnostically helpful when the differential diagnosis includes both small lymphocyte-like PCM and LPL, as the latter are almost invariably IgM positive whereas IgM-expressing PCMs are rare. Finally, HC/LC immunohistochemistry can aid in the characterization of hyposecretory or nonsecretory myelomas.

Interpretation of LC IHC is often difficult due to high background staining resulting from high serum paraprotein levels or because of suboptimal immunohistochemical staining technique. Chromogenic LC in situ hybridization (ISH) largely circumvents these technical issues [80]. LC ISH has virtually no background staining and has sensitivity comparable to that of LC IHC (Fig. 3.27). The results of LC ISH studies performed on harshly decalcified specimens (e.g., femoral heads) should be interpreted with caution, however, as such decalcification may yield a false negative result secondary to RNA degradation. In our experience, ISH can be successfully performed on bone marrow biopsies decalcified with milder organic acid-based reagents, which often include EDTA. ISH is usually incompatible with the harsh inorganic acid-containing solutions typically used for femoral heads and other bone specimens. For larger specimens when a PC

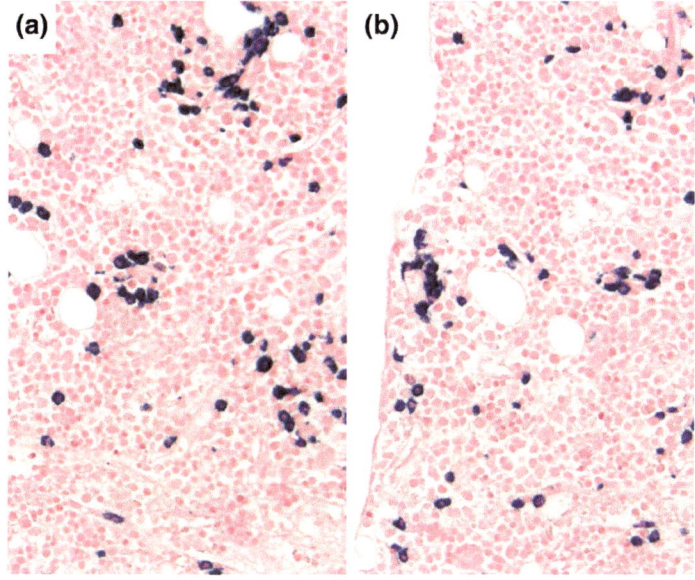

Fig. 3.27 In situ hybridization for kappa (**a**) and lambda (**b**) immunoglobulin light chain in a reactive bone marrow plasmacytosis. Polytypic light chain expression by the PCs is readily apparent given the virtual absence of background staining that is often present in light chain immunohistochemical stains

neoplasm is suspected, separate processing of curetted marrow, which may need little or no decalcification, provides tissue suitable for potential ISH studies.

CD20

Expression of CD20 is rapidly lost during B-cell ontogeny commensurate with the commitment to plasmacytic differentiation; consequently, normal PCs are CD20 negative. However, aberrant CD20 expression is detected in approximately 15% of PCMs and is strongly associated with the presence of the t(11;14) [81–85]. In contrast to benign B-cells and most malignant B-cell lymphomas, CD20 expression in PCM is typically heterogeneous and may only be evident in a subset of neoplastic PCs.

PAX5

PAX5 is a B-cell transcription factor whose expression is downregulated at the onset of plasmacytic differentiation and is thus not expressed in normal PCs [86, 87].

However, aberrant PAX5 expression is detected in a small subset of PCM. As with CD20, PAX5 positive PCM is associated with the t(11;14) and small lymphocyte-like morphology [31]. In addition, PAX5 positive PCM may express surface light chain. Because these PCMs share morphologic and immunophenotypic features of both mature B-cells and PCs, we have designated them as "transitional" PCM. Not surprisingly, transitional PCM is particularly prone to misdiagnosis and highlights the importance of cyclin D1 IHC and t(11;14) FISH in the distinction of such cases from LPL and other small B-cell lymphomas.

CD56

CD56 (neural cell adhesion molecule or NCAM) is aberrantly expressed in approximately 75% of PCM [81, 82, 88–90]. Because it is highly overexpressed in most cases, CD56 is readily detected by IHC. Within the spectrum of neoplasms with plasmacytic differentiation, CD56 expression occurs in LPL and other B-cell lymphomas with plasmacytic differentiation in addition to PCM; however, some studies indicate the frequency of CD56 positivity is significantly lower in these B-cell lymphomas than in PCM [58–60]. Although useful for evaluation of PCM, it is important to note that other bone marrow residents, such as osteoblasts and NK cells, normally express CD56 [91, 92].

Cyclin D1

Cyclin D1 is overexpressed in approximately 20% of PCM. Cyclin D1 IHC is particularly helpful when the differential diagnosis includes both low-grade B-cell lymphomas with plasmacytic differentiation and small lymphocyte-like PCM. The latter is strongly associated with the t(11;14) and is strongly cyclin D1 positive, whereas the former (e.g., LPL, MALT lymphoma) are cyclin D1 negative. It should also be noted that weak cyclin D1 positivity may occur in PCMs containing extra copies of the *CCND1* locus due to trisomy or tetrasomy of chromosome 11 but lacking the t(11;14) rearrangement [84].

References

1. Swerdlow SH, Campo E, Harris NL, et al. WHO classification of tumours of haematopoietic and lymphoid tissues. Lyon: IARC Press; 2008.
2. Dimopoulos MA, Pouli A, Anagnostopoulos A, et al. Macrofocal multiple myeloma in young patients: a distinct entity with favorable prognosis. Leuk Lymphoma. 2006;47:1553–6.
3. Walker R, Barlogie B, Haessler J, et al. Magnetic resonance imaging in multiple myeloma: diagnostic and clinical implications. J Clin oncol: Off J Am Soc Clin Oncol. 2007;25:1121–8.

4. Foucar K, Raber M, Foucar E, Barlogie B, Sandler CM, Alexanian R. Anaplastic myeloma with massive extramedullary involvement. Report of two cases. Cancer. 1983;51:166–74.
5. Khayyata S, Bentley G, Fregene TA, Al-Abbadi M. Retroperitoneal extramedullary anaplastic plasmacytoma masquerading as sarcoma: Report of a case with an unusual presentation and imprint smears. Acta Cytol. 2007;51:434–6.
6. Rao S, Kar R, Pati HP. Anaplastic myeloma: a morphologic diagnostic dilemma. Indian J Hematol Blood Transfus: Off J Indian Soc Hematol Blood Transfus. 2008;24:188–9.
7. Subitha K, Renu T, Lillykutty P, Letha V. Anaplastic myeloma presenting as mandibular swelling: Diagnosis by cytology. J Cytol/Indian Acad Cytol. 2014;31:114–6.
8. Kumar G, Ahluwalia J. "Multiple forms" of a myeloma. Blood. 2015;126:692.
9. Eyden BP, Banerjee SS. Multiple myeloma showing signet-ring cell change. Histopathology. 1990;17:170–2.
10. Dorfman RF. Multiple myeloma showing signet-ring cell change. Histopathology. 1991;18:577–8.
11. Haidar JH, Bazarbachi A, Nasr MR, El-Sabban ME, Daher R. Signet ring-like light chain myeloma with systemic spread. Eur J Haematol. 2003;70:249–50.
12. Grier DD, Robbins K. Signet-ring plasma cell myeloma. Am J Hematol. 2012;87:625.
13. Parmentier S, Radke J. Pseudo-Auer rods in a patient with newly diagnosed IgG myeloma. Blood. 2012;119:650.
14. Metzgeroth G, Back W, Maywald O, et al. Auer rod-like inclusions in multiple myeloma. Ann Hematol. 2003;82:57–60.
15. Al Muslahi M, Teague M, Lee SH, Roberts M. Multiple myeloma simulating Gaucher's disease. Br J Haematol 2006;134:123.
16. Brunning RD, Parkin J. Intranuclear inclusions in plasma cells and lymphocytes from patients with monoclonal gammopathies. Am J Clin Pathol. 1976;66:10–21.
17. Eyre TA, Littlewood TJ, Bain BJ. Dutcher bodies: cytoplasmic inclusions within the nucleus. Br J Haematol. 2014;166:946–7.
18. Jiang N, Qi C, Chang H. Dutcher bodies in multiple myeloma are highly associated with translocation t(4;14) and IgA isotype. Br J Haematol. 2015.
19. Maldonado JE, Bayrd ED, Brown AL Jr. The flaming cell in multiple myeloma. A light and electron microscopy study. Am J Clin Pathol. 1965;44:605–12.
20. Waldenstrom J, Paraskevas F, Heremans J. The incidence and cytology of different myeloma types. Lancet 1961;1:1147.
21. Savage DG, Zipin D, Bhagat G, Alobeid B. Hemophagocytic, non-secretory multiple myeloma. Leuk Lymphoma. 2004;45:1061–4.
22. Ramos J, Lorsbach R. Hemophagocytosis by neoplastic plasma cells in multiple myeloma. Blood. 2014;123:1634.
23. Galeotti J, Wu B, Wang E. Erythrophagocytosis in a cyclin D1 positive plasma cell myeloma with near-tetraploid karyotypic abnormalities and cryptic MYC/IGH fusion. Ann Hematol 2015.
24. Machaczka M, Vaktnas J, Klimkowska M, Nahi H, Hagglund H. Acquired hemophagocytic lymphohistiocytosis associated with multiple myeloma. Med Oncol. (Northwood, London, England) 2011;28:539–43.
25. Greipp PR, Raymond NM, Kyle RA, O'Fallon WM. Multiple myeloma: significance of plasmablastic subtype in morphological classification. Blood. 1985;65:305–10.
26. Greipp PR, Leong T, Bennett JM, et al. Plasmablastic morphology–an independent prognostic factor with clinical and laboratory correlates: Eastern Cooperative Oncology Group (ECOG) myeloma trial E9486 report by the ECOG Myeloma Laboratory Group. Blood. 1998;91:2501–7.
27. Bartl R, Frisch B, Burkhardt R, et al. Bone marrow histology in myeloma: its importance in diagnosis, prognosis, classification and staging. Br J Haematol. 1982;51:361–75.
28. Bartl R, Frisch B, Fateh-Moghadam A, Kettner G, Jaeger K, Sommerfeld W. Histologic classification and staging of multiple myeloma. A retrospective and prospective study of 674 cases. Am J Clin Pathol. 1987;87:342–55.

29. Hoyer JD, Hanson CA, Fonseca R, Greipp PR, Dewald GW, Kurtin PJ. The (11;14)(q13;q32) translocation in multiple myeloma. A morphologic and immunohistochemical study. Am J Clin Pathol. 2000;113:831–7.
30. Garand R, vet-Loiseau H, Accard F, Moreau P, Harousseau JL, Bataille R. t(11;14) and t(4;14) translocations correlated with mature lymphoplasmacytoid and immature morphology, respectively, in multiple myeloma. Leukemia. 2003;17:2032–5.
31. Lin P, Mahdavy M, Zhan F, Zhang HZ, Katz RL, Shaughnessy JD. Expression of PAX5 in CD20-positive multiple myeloma assessed by immunohistochemistry and oligonucleotide microarray. Mod Pathol: Off J US Can Acad Pathol, Inc 2004;17:1217–22.
32. Alapat D, Viswanatha D, Xie M, Lorsbach R. Plasma cell myeloma (PCM) with immunophenotypic features transitional between that of myeloma and lymphoma. Mod Pathol: Off J US Can Acad Pathol, Inc 2012;25:321A.
33. Lin P, Hao S, Handy BC, Bueso-Ramos CE, Medeiros LJ. Lymphoid neoplasms associated with IgM paraprotein: a study of 382 patients. Am J Clin Pathol. 2005;123:200–5.
34. Jenner E. Serum free light chains in clinical laboratory diagnostics. Clin Chim Acta. 2013;427:15–20.
35. Drayson M, Tang LX, Drew R, Mead GP, Carr-Smith H, Bradwell AR. Serum free light-chain measurements for identifying and monitoring patients with nonsecretory multiple myeloma. Blood. 2001;97:2900–2.
36. Avet-Loiseau H, Garand R, Lode L, Harousseau JL, Bataille R. Translocation t(11;14)(q13; q32) is the hallmark of IgM, IgE, and nonsecretory multiple myeloma variants. Blood. 2003;101:1570–1.
37. Ramos J, Alapat D, Lorsbach RB. Characterization of non-secretory plasma cell myeloma. Manuscript in preparation 2015.
38. Cogne M, Guglielmi P. Exon skipping without splice site mutation accounting for abnormal immunoglobulin chains in nonsecretory human myeloma. Eur J Immunol. 1993;23:1289–93.
39. Coriu D, Weaver K, Schell M, et al. A molecular basis for nonsecretory myeloma. Blood. 2004;104:829–31.
40. Ramsingh G, Mehan P, Luo J, Vij R, Morgensztern D. Primary plasma cell leukemia: a surveillance, epidemiology, and end results database analysis between 1973 and 2004. Cancer. 2009;115:5734–9.
41. Fernandez de Larrea C, Kyle RA, Durie BG, et al. Plasma cell leukemia: consensus statement on diagnostic requirements, response criteria and treatment recommendations by the International Myeloma Working Group. Leukemia. 2013;27:780–91.
42. Garcia-Sanz R, Orfao A, Gonzalez M, et al. Primary plasma cell leukemia: clinical, immunophenotypic, DNA ploidy, and cytogenetic characteristics. Blood. 1999;93:1032–7.
43. Tiedemann RE, Gonzalez-Paz N, Kyle RA, et al. Genetic aberrations and survival in plasma cell leukemia. Leukemia. 2008;22:1044–52.
44. Kyle RA, Gertz MA, Witzig TE, et al. Review of 1027 patients with newly diagnosed multiple myeloma. Mayo Clin Proc. 2003;78:21–33.
45. Merlini G, Stone MJ. Dangerous small B-cell clones. Blood. 2006;108:2520–30.
46. Vrana JA, Gamez JD, Madden BJ, Theis JD, Bergen HR III, Dogan A. Classification of amyloidosis by laser microdissection and mass spectrometry-based proteomic analysis in clinical biopsy specimens. Blood. 2009;114:4957–9.
47. Desikan KR, Dhodapkar MV, Hough A, et al. Incidence and impact of light chain associated (AL) amyloidosis on the prognosis of patients with multiple myeloma treated with autologous transplantation. Leuk Lymphoma. 1997;27:315–9.
48. Bahlis NJ, Lazarus HM. Multiple myeloma-associated AL amyloidosis: is a distinctive therapeutic approach warranted? Bone Marrow Transplant. 2006;38:7–15.
49. Jones D, Bhatia VK, Krausz T, Pinkus GS. Crystal-storing histiocytosis: a disorder occurring in plasmacytic tumors expressing immunoglobulin kappa light chain. Hum Pathol. 1999;30:1441–8.

50. Lebeau A, Zeindl-Eberhart E, Muller EC, et al. Generalized crystal-storing histiocytosis associated with monoclonal gammopathy: molecular analysis of a disorder with rapid clinical course and review of the literature. Blood. 2002;100:1817–27.
51. Ionescu DN, Pierson DM, Qing G, Li M, Colby TV, Leslie KO. Pulmonary crystal-storing histiocytoma. Arch Pathol Lab Med. 2005;129:1159–63.
52. Abildgaard N, Bendix-Hansen K, Kristensen JE, et al. Bone marrow fibrosis and disease activity in multiple myeloma monitored by the aminoterminal propeptide of procollagen III in serum. Br J Haematol. 1997;99:641–8.
53. Subramanian R, Basu D, Dutta TK. Significance of bone marrow fibrosis in multiple myeloma. Pathology. 2007;39:512–5.
54. Babarovic E, Valkovic T, Stifter S, et al. Assessment of bone marrow fibrosis and angiogenesis in monitoring patients with multiple myeloma. Am J Clin Pathol. 2012;137:870–8.
55. Brouet JC, Fermand JP, Laurent G, et al. The association of chronic lymphocytic leukaemia and multiple myeloma: a study of eleven patients. Br J Haematol. 1985;59:55–66.
56. Chang H, Wechalekar A, Li L, Reece D. Molecular cytogenetic abnormalities in patients with concurrent chronic lymphocytic leukemia and multiple myeloma shown by interphase fluorescence in situ hybridization: evidence of distinct clonal origin. Cancer Genet Cytogenet. 2004;148:44–8.
57. Kaufmann H, Ackermann J, Nosslinger T, et al. Absence of clonal chromosomal relationship between concomitant B-CLL and multiple myeloma—a report on two cases. Ann Hematol. 2001;80:474–8.
58. Seegmiller AC, Xu Y, McKenna RW, Karandikar NJ. Immunophenotypic differentiation between neoplastic plasma cells in mature B-cell lymphoma vs plasma cell myeloma. Am J Clin Pathol. 2007;127:176–81.
59. Morice WG, Chen D, Kurtin PJ, Hanson CA, McPhail ED. Novel immunophenotypic features of marrow lymphoplasmacytic lymphoma and correlation with Waldenstrom's macroglobulinemia. Mod Pathol. 2009;22:807–16.
60. Rosado FG, Morice WG, He R, Howard MT, Timm M, McPhail ED. Immunophenotypic features by multiparameter flow cytometry can help distinguish low grade B-cell lymphomas with plasmacytic differentiation from plasma cell proliferative disorders with an unrelated clonal B-cell process. Br J Haematol. 2015;169:368–76.
61. Hagen W, Schwarzmeier J, Walchshofer S, et al. A case of bone marrow mastocytosis associated with multiple myeloma. Ann Hematol. 1998;76:167–74.
62. Motwani P, Kocoglu M, Lorsbach RB. Systemic mastocytosis in association with plasma cell dyscrasias: report of a case and review of the literature. Leuk Res. 2009;33:856–9.
63. Stellmacher F, Sotlar K, Balleisen L, Valent P, Horny HP. Bone marrow mastocytosis associated with IgM kappa plasma cell myeloma. Leuk Lymphoma. 2004;45:801–5.
64. Chilosi M, Adami F, Lestani M, et al. CD138/syndecan-1: a useful immunohistochemical marker of normal and neoplastic plasma cells on routine trephine bone marrow biopsies. Mod Pathol. 1999;12:1101–6.
65. Costes V, Magen V, Legouffe E, et al. The Mi15 monoclonal antibody (anti-syndecan-1) is a reliable marker for quantifying plasma cells in paraffin-embedded bone marrow biopsy specimens. Hum Pathol. 1999;30:1405–11.
66. O'Connell FP, Pinkus JL, Pinkus GS. CD138 (syndecan-1), a plasma cell marker immunohistochemical profile in hematopoietic and nonhematopoietic neoplasms. Am J Clin Pathol. 2004;121:254–63.
67. Al-Quran SZ, Yang L, Magill JM, Braylan RC, Douglas-Nikitin VK. Assessment of bone marrow plasma cell infiltrates in multiple myeloma: the added value of CD138 immunohistochemistry. Hum Pathol. 2007;38:1779–87.
68. Ng AP, Wei A, Bhurani D, Chapple P, Feleppa F, Juneja S. The sensitivity of CD138 immunostaining of bone marrow trephine specimens for quantifying marrow involvement in MGUS and myeloma, including samples with a low percentage of plasma cells. Haematologica. 2006;91:972–5.

69. Chu PG, Arber DA, Weiss LM. Expression of T/NK-cell and plasma cell antigens in nonhematopoietic epithelioid neoplasms. An immunohistochemical study of 447 cases. Am J Clin Pathol. 2003;120:64–70.
70. Kambham N, Kong C, Longacre TA, Natkunam Y. Utility of syndecan-1 (CD138) expression in the diagnosis of undifferentiated malignant neoplasms: a tissue microarray study of 1,754 cases. Appl Immunohistochem Mol Morphol. 2005;13:304–10.
71. Bayer-Garner IB, Sanderson RD, Dhodapkar MV, Owens RB, Wilson CS. Syndecan-1 (CD138) immunoreactivity in bone marrow biopsies of multiple myeloma: shed syndecan-1 accumulates in fibrotic regions. Mod Pathol. 2001;14:1052–8.
72. Mahtouk K, Hose D, Raynaud P, et al. Heparanase influences expression and shedding of syndecan-1, and its expression by the bone marrow environment is a bad prognostic factor in multiple myeloma. Blood. 2007;109:4914–23.
73. Shaffer AL, Emre NC, Lamy L, et al. IRF4 addiction in multiple myeloma. Nature. 2008;454:226–31.
74. Falini B, Fizzotti M, Pucciarini A, et al. A monoclonal antibody (MUM1p) detects expression of the MUM1/IRF4 protein in a subset of germinal center B cells, plasma cells, and activated T cells. Blood. 2000;95:2084–92.
75. Natkunam Y, Warnke RA, Montgomery K, Falini B, Van de RM. Analysis of MUM1/IRF4 protein expression using tissue microarrays and immunohistochemistry. Mod Pathol. 2001;14:686–94.
76. Klopfenstein DR, Klumperman J, Lustig A, Kammerer RA, Oorschot V, Hauri HP. Subdomain-specific localization of CLIMP-63 (p63) in the endoplasmic reticulum is mediated by its luminal alpha-helical segment. J Cell Biol. 2001;153:1287–300.
77. Banham AH, Turley H, Pulford K, Gatter K, Mason DY. The plasma cell associated antigen detectable by antibody VS38 is the p63 rough endoplasmic reticulum protein. J Clin Pathol. 1997;50:485–9.
78. Turley H, Jones M, Erber W, Mayne K, de WM, Gatter K. VS38: a new monoclonal antibody for detecting plasma cell differentiation in routine sections. J Clin Pathol 1994;47:418–22.
79. Shanks JH, Banerjee SS. VS38 immunostaining in melanocytic lesions. J Clin Pathol. 1996;49:205–7.
80. Beck RC, Tubbs RR, Hussein M, Pettay J, Hsi ED. Automated colorimetric in situ hybridization (CISH) detection of immunoglobulin (Ig) light chain mRNA expression in plasma cell (PC) dyscrasias and non-Hodgkin lymphoma. Diagn Mol Pathol: Am J Surg Pathol, part B. 2003;12:14–20.
81. Lin P, Owens R, Tricot G, Wilson CS. Flow cytometric immunophenotypic analysis of 306 cases of multiple myeloma. Am J Clin Pathol. 2004;121:482–8.
82. Mateo G, Montalban MA, Vidriales MB, et al. Prognostic value of immunophenotyping in multiple myeloma: a study by the PETHEMA/GEM cooperative study groups on patients uniformly treated with high-dose therapy. J Clin Oncol. 2008;26:2737–44.
83. Robillard N, vet-Loiseau H, Garand R, et al. CD20 is associated with a small mature plasma cell morphology and t(11;14) in multiple myeloma. Blood 2003;102:1070–1.
84. Cook JR, Hsi ED, Worley S, Tubbs RR, Hussein M. Immunohistochemical analysis identifies two cyclin D1+ subsets of plasma cell myeloma, each associated with favorable survival. Am J Clin Pathol. 2006;125:615–24.
85. Heerema-McKenney A, Waldron J, Hughes S, et al. Clinical, immunophenotypic, and genetic characterization of small lymphocyte-like plasma cell myeloma: a potential mimic of mature B-cell lymphoma. Am J Clin Pathol. 2010;133:265–70.
86. Torlakovic E, Torlakovic G, Nguyen PL, Brunning RD, Delabie J. The value of anti-pax-5 immunostaining in routinely fixed and paraffin-embedded sections: a novel pan pre-B and B-cell marker. Am J Surg Pathol. 2002;26:1343–50.
87. Feldman AL, Dogan A. Diagnostic uses of Pax5 immunohistochemistry. Adv Anat Pathol. 2007;14:323–34.

88. Almeida J, Orfao A, Ocqueteau M, et al. High-sensitive immunophenotyping and DNA ploidy studies for the investigation of minimal residual disease in multiple myeloma. Br J Haematol. 1999;107:121–31.
89. Van CB, Durie BG, Spier C, et al. Plasma cells in multiple myeloma express a natural killer cell-associated antigen: CD56 (NKH-1; Leu-19). Blood. 1990;76:377–82.
90. Garcia-Sanz R, Gonzalez M, Orfao A, et al. Analysis of natural killer-associated antigens in peripheral blood and bone marrow of multiple myeloma patients and prognostic implications. Br J Haematol. 1996;93:81–8.
91. Ely SA, Knowles DM. Expression of CD56/neural cell adhesion molecule correlates with the presence of lytic bone lesions in multiple myeloma and distinguishes myeloma from monoclonal gammopathy of undetermined significance and lymphomas with plasmacytoid differentiation. Am J Pathol. 2002;160:1293–9.
92. Morice WG. The immunophenotypic attributes of NK cells and NK-cell lineage lymphoproliferative disorders. Am J Clin Pathol. 2007;127:881–6.

Chapter 4
Cytogenetics of Plasma Cell Neoplasms

Jeffrey R. Sawyer

Introduction

Plasma cell (PC) neoplasms are clonal bone marrow diseases characterized by the transformation of differentiated B-cells. They are heterogeneous diseases both at the genetic level and in terms of clinical outcome [1, 2]. Advances in molecular genetics and genomic profiling studies have provided a greater understanding of the genetic pathogenesis of myeloma and provide the rationale for the current molecular sub-classifications of the disease [3–9]. This chapter describes the clonal progression of both primary structural and numerical chromosomal aberrations and how the sequential acquisition of secondary chromosomal aberrations leads to cytogenetically defined high-risk disease. Risk stratification by interphase fluorescence in situ hybridization (iFISH) in combination with metaphase cytogenetics provides the greatest understanding of the underlying numerical and structural aberrations which occur in PC disorders.

In newly diagnosed patients, the abnormal clone will usually have a low proliferative activity; as a result the metaphase cells identified are in many cases derived from normal hematopoiesis. Therefore, only about 30–40% of myeloma patients will show an abnormal metaphase karyotype [10, 11]. The detection of an abnormal metaphase karyotype is generally correlated with an elevated plasma cell labeling index and tumor burden, which is reflected in a higher mitotic rate [12]. Therefore, finding abnormal mitoses in a sample can, in a certain respect, be regarded as a surrogate marker for stroma-independent cells and more advanced

J.R. Sawyer (✉)
Department of Pathology and Myeloma Institute, University of Arkansas for Medical
Sciences, Little Rock, AR 72205, USA
e-mail: sawyerjeffreyr@uams.edu

J.R. Sawyer
Cytogenetics Laboratory, Freeway Medical Tower, Suite 200, 5800 West 10th Street, Little
Rock, AR 72204, USA

© Springer International Publishing Switzerland 2016
R.B. Lorsbach and M. Yared (eds.), *Plasma Cell Neoplasms*,
DOI 10.1007/978-3-319-42370-8_4

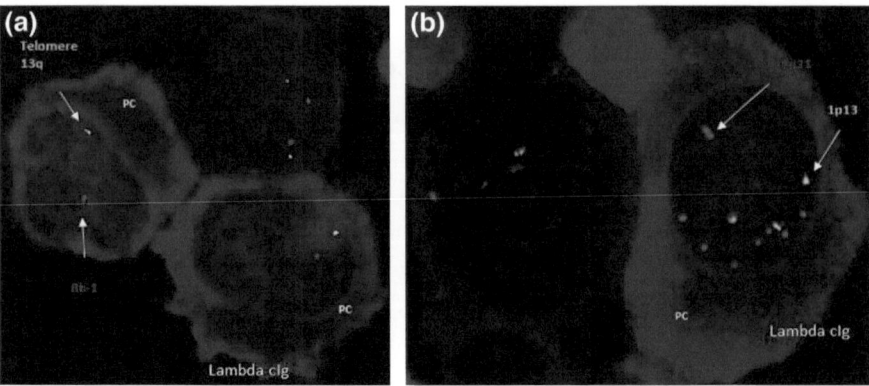

Fig. 4.1 Interphase FISH of bone marrow plasma cells (PC) of patients with MM. **a** A single copy of FISH probes for both the RB1 locus (*red*, 13q12), and the 13q telomere probe (*green*) indicate monsomy 13 in these cells. Cytoplasm of plasma cells is stained *blue* with a lambda cytoplasmic immunoglobulin (cIg). **b** Multiple copies of 1q21 (*red*) and single copies of 1p13 probe (*green*) indicate amplification of 1q21 in these cells. iFISH images courtesy Dr. Erming Tian

disease [13]. The low proliferative activity of the tumor cells in the early stages of the disease is the main limitation of metaphase cytogenetics, since only dividing cells can be analyzed. Even within cells with an abnormal karyotype some aberrations can be cryptic in metaphase analysis; for example the t(4;14)(p16;q32) is a submicroscopic aberration and cannot be detected with conventional chromosome banding techniques. Many of the limitations of chromosome banding techniques have been overcome by the use of iFISH (Fig. 4.1), locus specific metaphase FISH (Fig. 4.2), and Spectral Karyotyping (SKY) (Fig. 4.3). iFISH analysis of interphase nuclei is independent of plasma cell division and as a result is the primary test used in cytogenetic stratification of new patients. It is currently standard practice that iFISH studies be performed as part of the diagnostic workup; however, just as metaphase analysis has limitations, iFISH studies also have certain drawbacks. Unfortunately, most iFISH panels contain only a few probes and are limited to detecting only these specific aberrations. Any structural or numerical aberration involving an uncommon aneuploidy or complex structural rearrangement of chromosomes not in the probe panel will not be detected. Several studies have reinforced the importance of both conventional cytogenetics and FISH, both of which are independent prognosticators at diagnosis [14–16].

Numerical Chromosome Aberrations

Chromosome aberrations in MM are typically complex and represent a hallmark of the disease. In most cases, many chromosomes are altered both numerically and structurally with a median number of eight karyotypic changes per patient [17].

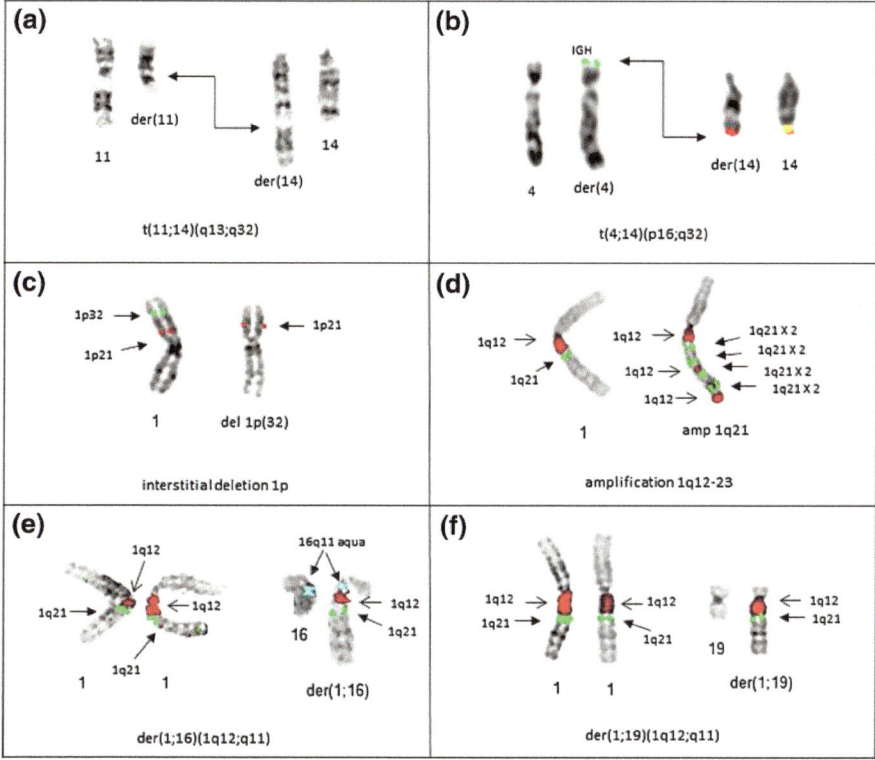

Fig. 4.2 Partial karyotypes of metaphase chromosomes depicting primary and secondary chromosome aberrations. **a** The t(11;14)(q13;q32) is the most common reciprocal translocation in MM, shown by G-banding. **b** Metaphase FISH of the t(4;14)(p16;q32) is shown by inverse DAPI G-bands. The IGH probe (*green*) has translocated from the 14q32 to the der(4) at 4p16. **c** Metaphase FISH showing interstitial deletion of 1p by inverse DAPI banding. Note presence of probes for 1p12 (*red*) and 1p32 (*green*) on normal chromosome 1 on *left* and absence of *green* signal 1p32 in del(1p32) chromosome on right. **d** Amplification of 1q12 ∼ 23 by breakage-fusion bridge (BFB) cycles. Note three copies of pericentromeric heterochromatin (*red, open arrows*) bracketing eight copies of 1q21 (*green, solid arrows*) on the *right*. **e** Whole-arm translocation of 1q12 to 16q11 resulting in three copies of 1q21 (*green, solid arrows*) and deletion of 16q. Chromosome 1 pericentromeric region 1q12 (*red, open arrows*) and 16 pericentromeric region 16q11 (aqua, *solid arrow*) are fused in an unbalanced whole-arm der(1;16)(q12;q11). **f** Whole-arm translocation of 1q12 (*red, open arrows*) to 19q11 resulting in three copies of 1q21 (*green, solid arrows*) and deletion of 19q in the der(1;19)

Statistical analysis of the progression of the hyperdiploid karyotypes indicates that most tumors have as many as ten abnormalities at diagnosis, predominately numerical trisomies. These clones can undergo expansion, generating subclones that are heterogeneous with subsequent secondary translocation events [17, 18]. In many ways, the complexity of karyotypes in MM resembles those found in solid tumors, because they combine high numbers of both numerical aberrations and very complex and unstable structural aberrations.

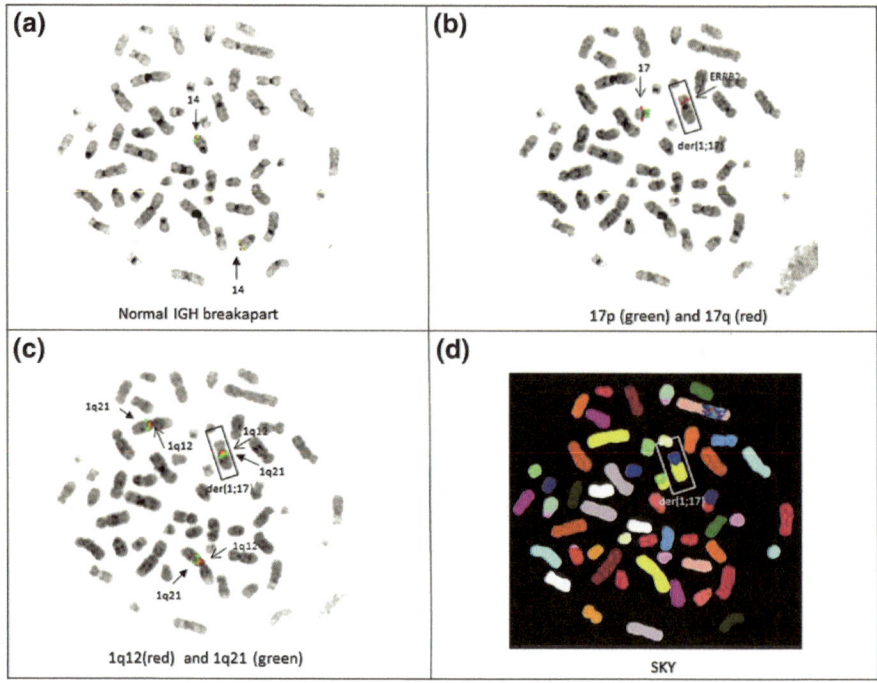

Fig. 4.3 A metaphase cell demonstrating the accumulation of high-risk secondary aberrations (del17p and 1q21 gain) in a low-risk progenitor clone with hyperdiploidy (54 chromosomes) and no IGH translocation. This cell was rehybridized with three different sets of locus specific FISH probe combinations (**a–c**) and Spectral Karyotyping (**d**). **a** IGH break apart probe shows chromosomes 14 with both proximal (*red*) and distal (*green*) signals remaining on the chromosomes 14 (*arrows*), indicating no translocation. **b** Normal signal pattern for chromosome 17 exhibits p53 on the short arm (green), and ERBB2 on the long arm (*red*) (*open arrow*). The der (1;17) inside box shows ERBB2 on 17q and the deletion of 17p resulting from a whole-arm translocation of 1q12 to 17p. **c** FISH probes for 1q12 (*red, open arrows*) and 1q21 (*green, solid arrows*) show three copies of 1q21. The der(1;17) inside box shows 1q12 and 1q21 on the der(17). **d** Sky demonstrates hyperdiploidy with common trisomies of chromosomes 5, 9, 15, 19, and the whole-arm translocation der(1;17) inside box (17q in *blue* and 1q in *yellow*)

The first prognostically significant cytogenetic classification of MM was based on chromosome number of the abnormal clones [19]. Conventional karyotyping identified two karyotypic subgroups: a hyperdiploid group with 51–55 chromosomes, and a non-hyperdipliod (hypodiploid) group with 43–46 chromosomes [19, 20]. The hyperdiploid clones are characterized by the gain of a distinct set of trisomies of chromosomes 3, 5, 7, 9, 11, 15, 19, and 21 (Table 4.1), which is observed in 50–60% of patients [10, 11, 17, 19–22]. The non-hyperdiploid myeloma is composed of the hypodiploid, pseudodiploid, and near-tetraploid variants and is almost always associated with structural aberrations. The near tetraploid karyotypes appear to be 4n duplications of cells having pseudodiploid or hypodiploid karyotypes [19, 21]. The most frequently lost chromosomes in

Table 4.1 Recurring characteristic chromosome aberrations in plasma cell disorders

Numerical aberrations	Frequency (%)	Chromosomes
Hyperdiploidy	50%	3, 5, 7, 9, 11, 15, 19, 21
Hypodiploidy	40%	−13, −14, −16, −22
Translocations		**Involved genes**
t(11;14)(q13;32)	25–30	cyclin D1/IgH
t(4;14)(p16.3;q32)	20–25	FGFR3/IgH
t(14;16)(q32;q22-23)	20–25	IgH/MAF
t(14;20)(q32;q12)	<5	IgH/MAFB
t(6;14)(p25;q32)	<5	IRF4/IgH
t(9;14)(p13;q32)	<5	PAX5/IgH

non-hyperdiploid clones are chromosomes 13, 14, 16 and 22 [21]. The non-hyperdiploid karyotypes typically have structural aberrations, usually involving translocations of the immunoglobulin heavy chain (IGH) locus at 14q32. The hypodiploid group is associated with poorer overall survival, while the hyperdiploid group does better [20]. These two ploidy-based subtypes of MM have been confirmed by FISH and molecular genetic analysis. It should be pointed out that the generalized distinctions between the hyperdiploid and the hypodiploid groups are an oversimplification since the primary IGH translocations are also found in hyperdiploid MM at a frequency of approximately 10% [23].

Primary Structural Aberrations in PC Disorders

It is known that primary IGH translocations occur early in the pathogenesis of MM whereas secondary translocations occur later and are involved in tumor progression [1]. Most primary translocations are simple reciprocal translocations and juxtapose an oncogene and one of the immunoglobulin enhancers. These translocations are mediated by errors in one of three B-cell specific DNA modification mechanisms, mostly IGH switch recombination, errors in somatic hypermutation, and rarely, VDJ recombination [24, 25]. The incidence of IGH translocations is known to increase during the progression of the disease from 50% in MGUS to 90% in human myeloma cell lines (HMCL). The incidence of light-chain translocations (kappa, 2p12 or lambda, 22q11) is far less, with about 10% being found in MGUS and 20% in advanced myeloma and HMCL [24–29].

The IGH (14q32) translocations found in non-hyperdiploid MM are promiscuous and can involve many different translocation partners. Typically, however, there are five main translocations involving 11q13 (*CCND1*) (Fig. 4.2a), 4p16 (FGFR3 and MMSET) (Fig. 4.2b), 6p21 (*CCND3*), 16q23 (*MAF*), and 20q12 (*MAFB*). Virtually all MM and MGUS tumors have Cyclin D dysregulation, providing an early and unifying pathogenetic event in MM [26]. The most frequent translocation in MM is the t(11;14)(q13;q32), which is found in about 15% of patients. This translocation appears to be associated with a favorable outcome in most series but can also appear in patients with an aggressive phenotype such as plasma cell leukemia. Therefore, the t(11;14) is regarded as neutral with regard to prognosis [21].

The second most frequent reciprocal translocation in MM is t(4;14)(p16;q32) which occurs in about 10% of patients. This translocation involves two proteins that are encoded on 4p16: the Wolf-Hirschhorn syndrome candidate 1 gene (WHSC1), also known as multiple myeloma SET domain [MMSET], a protein with homology to histone methyltransferases; and fibroblast growth factor receptor 3 (FGRR3), an oncogenic receptor tyrosine kinase [1]. The translocation is unbalanced in up to 25% of cases, with the loss of the derivative chromosome 14 which is associated with the loss of FGFR3 expression [30]. As previously noted, the t(4;14)(p16;q32) is cryptic to banding and only detectable by FISH or reverse transcriptase PCR. The t(4;14) translocation is also associated with a high prevalence of chromosome 13 monosomy and deletions, and has been associated with poor survival [7, 30–32]. The t(4;14)(p16;q32) is also observed in MGUS but is present more frequently in patients with smoldering myeloma (SM) and active myeloma [21, 33].

Two less frequent but clinically important translocations involve the MAF genes. The t(14;16)(q32;q23) juxtaposes the IGH locus (14q32) and c-*MAF* locus (16q23) and is found in 6–7% of patients with MM [34]. The breakpoints occur in the introns of WWOX, a very large gene that spans a fragile site (FRA16D) on chromosome 16q23 [1]. This translocation is also associated with poor outcome [21]. Even less frequent is the MAFB (20q12) translocation which occurs in about 2% of patients and involves the reciprocal t(14;20)(q32;q12). The prognostic outcome is assumed to be the same as the t(14;16). It seems unlikely that any of the IGH translocations alone are sufficient to give rise to symptomatic myeloma as they are often evenly distributed between MGUS, SMM and MM [35].

It is known that these same independent IGH translocations can and do occur together, indicating the possible activation of complimentary pathways by multiple primary translocations [1]. These combinations of translocations include cyclin D3 and c-*MAF*, c-*MAF* and FGFR3/MMSET; and FGRF3/MMSET and cyclin D1. These combinations are far more likely to be found in late disease where complex rearrangements also involve a large number of other secondary aberrations.

Cytogenetic Aberrations Present in MGUS

PC neoplasms progress through stages which, in most cases, include a pre-malignant stage called monoclonal gammopathy of undetermined significance (MGUS) [1]. MGUS is a PC proliferative disorder with a bone marrow PC content of <10%, a monoclonal protein spike of <30 g/L, and no end organ damage. Routine metaphase cytogenetic analysis in MGUS is usually uninformative; hence iFISH studies are routinely used to identify cytogenetic aberrations. The primary IGH translocations originate at the MGUS stage of the disease (Table 4.1) and are felt to be causally related to the onset of clonal PC proliferation [36]. Aberrations identified by iFISH in the MGUS stage include the IGH translocations of t(11;14), t(4;14), t(14;16), all of the trisomic numerical aberrations, and monosomy/deletion 13 [21, 33, 37]. Although the primary IGH translocations and hyperdiploid numerical aberrations are commonly identified by iFISH in MGUS, the meaning of the aberrations is unclear, since the aberrations appear to be genetically dormant at this stage of the disease. In MGUS, IGH translocations are found in about 50% of patients and nearly half are hyperdiploid [38]. It may be that it is the proportion of PCs with specific genetic aberrations that accounts for the transition from MGUS to smoldering MM rather than the mere presence of the aberrations [39].

Cytogenetic Aberrations Present in Smoldering MM

Smoldering myeloma (SMM) appears to involve the continued clonal expansion of genetically and cytogenetically abnormal PCs [39]. Although IGH translocations and deletion 13 are found in MGUS, there is a significant increase in the proportion of PCs with IGH translocations and 13q deletions found in the transition from MGUS to SMM and from SMM to MM [39]. The Mayo Clinic has proposed a cytogenetically defined risk-based classification of smoldering MM that includes the aberrations of t(4;14), del(17p), and 1q gain for the high-risk group. Intermediate-risk aberrations include trisomies without IGH translocations. Standard-risk aberrations include other IGH translocations including t(11;14), t (14;16) and t(14;20), and the presence of trisomies and IGH translocations except for t(4;14) and monosomy13/del(13q). Low-risk disease has no detectable cytogenetic aberrations [36].

Cytogenetic Aberrations in Symptomatic Myeloma

Symptomatic myeloma involves all of the primary aberrations described for MGUS and SMM but is distinguished by the accumulation of a number of characteristic secondary aberrations. Although a very large number of secondary chromosomal aberrations are possible during disease progression, there are four main chromosome aberrations which are thought to be most clinically relevant to disease progression and prognosis. These aberrations include translocations and copy number (CN) aberrations of the *MYC* locus, deletions of chromosomes 17p and 1p, and the amplification of chromosome region 1q21 (Fig. 4.1b) [1]. Translocations and/or amplifications of the oncogene *MYC* (8q24) are involved in up to 45% of patients with advanced MM. These most common *MYC* translocations include the t(8;14) (q24;q32), t(2;8)(p12;q24), t(8;22)(q24;q12) and the t(1;8)(p12;q24) (Table 4.2).

Table 4.2 Secondary
chromosome aberrations

Translocations	Involved genes
t(8;14)(q24;q32)	MYC/IgH
t(2;8)(p12;q24)	IgK/MYC
t(8;22)(q24;q32)	MYC/IgL
t(1;8)(p12;q24)	FAM46C/MYC
Deletions	**Involved genes**
del(1)(p32)	CBKN2C
del(13)(q14~25)	Rb, DBM
del(17)(p13)	p53
del(8p)	–
del(6q)	–
del(16q)	CYLD,WWOX
1q21 Amplifications	–
amp1q21-23	MCL1,CKS1B
der(1;16)(q12;q11)	–
der(1;19)(q12;q11)	–
der(1;17)(q12;p11)	–

Translocations involving *MYC* and the IG locus are known to be late events in tumor progression when tumors are becoming more proliferative and less stromal dependent [25, 40]. The translocations that juxtapose *MYC* and IG sequences in myeloma are many times very complex events involving multiple chromosomes. They can involve non-reciprocal rearrangements, duplications, and amplifications and can be mediated by secondary translocations that do not involve B-cell specific modification mechanisms. Interphase FISH studies indicate that *MYC* rearrangements are present in 15% of myeloma tumors and are often heterogeneous within cells of the tumor. The secondary rearrangements involving IG and *MYC* loci show a similar prevalence in hyperdiploid and non-hyperdiploid myeloma, as do the *MYC* rearrangements that do not involve the IG heavy or light chain locus [40]. It has been reported that promiscuous translocations of 8q24 juxtapose *MYC* with genes that harbor super-enhancers, resulting in overexpression and an aggressive disease phenotype [41, 42]. A recurring reciprocal translocation of 1p, the t(1;8)(p12;q24) occurs as a secondary aberration and juxtaposes the *MYC* to 1p13. This is an example of a secondary *MYC* translocation that does not involve an IG rearrangement [40, 43, 44] and in some cases involves FAM46C at 1p12 [41].

Plasma cell leukemia (PCL) is an aggressive form of MM defined by peripheral blood involvement and may be present initially (primary PCL) or develop later in the disease course (secondary PCL). PCL reflects continued clonal evolution of cytogenetic aberrations found in myeloma [1, 2, 10, 11]. PCL is characterized by very complex karyotypes and genomic instability including frequent isochromosome 1q aberrations. Chromosome 1q aberrations are particularly complex and tend to become unstable during progression to extramedullary disease [10, 11].

Chromosome 13 Aberrations

Chromosome 13 aberrations are found in about 50% of cases, with most being complete monosomy 13 (85%) while the remaining 15% constitute deletion 13 (Fig. 4.1a). Historically, del(13q) has been associated with an unfavorable prognosis in MM; however, increasing evidence indicates that its prognostic relevance may be related to its association with other genetic aberrations. The first link between a recurrent chromosome abnormality and prognosis in MM was identified when monosomy and/or deletion 13 were associated with aggressive clinical course [14, 21, 45–48]. More recent studies suggest that the prognostic significance of monosomy 13 may emanate from its close association with the t(4;14)(p16;q32) translocation.

Deletions of 17p13

The deletion of 17p in MM leads to the loss of heterozygosity of *p53*, a well characterized tumor suppressor gene that transcriptionally regulates cell-cycle progression and apoptosis. The deletion or inactivation of *p53* is a rare late event, with deletions of 17p13 being reported in 10% of patients by iFISH [21]. Clinically it has been reported that patients with 17p deletions have more aggressive disease, a higher prevalence of extramedullary disease and overall shorter survival [7, 21, 49]. FISH studies of clonal evolution in myeloma indicate the deletion occurs most commonly in subclones [50].

A model for the clonal evolution of MM based on iFISH studies was proposed in 2005 by Cremer et al. [50] which delineates distinct subgroups of MM. In this model clustering analysis indicated four major branches of an oncogenetic tree including a hyperdiploid branch, a del(13) and t(4;14) branch, a t(11;14) branch, and a gain of 1q branch. Statistical modeling of this oncogenetic tree indicates early independent events were gains of 15, 9, 11 and t(11;14), deletion of 13q followed by t(4;14), and finally gain of 1q. Aberrations of 17p, 22q, 8p and 6q were subsequent events. The t(4;14) was associated with the non-hyperdiploid pathway and was always preceded by deletion 13q. Deletions of 6q, 8p, and 22q were linked to the major branch of non-hyperdiploid MM [50]. This model also indicates that the deletion of 17p13 is a late secondary event in MM. In the pathway to hyperdiploidy, gains of 9 and 15 were associated, while gain of 19 was a subsequent event. Interestingly, they found gain of 1q and 17p at approximately the same frequency in both the hyperdiploid and non-hyperdiploid groups, indicating the presence of a subgroup of hyperdiploid patients with adverse aberrations.

Chromosome 1p Deletions

Chromosome 1 aberrations are the most common structural aberrations in MM and mostly involve deletions in the 1p and amplifications of 1q12~23. The deletions of 1p are mainly interstitial and have been characterized by banding and FISH studies [10, 11, 17]. They are defined by losses spanning the 1p13–1p31 region (Fig. 4.2c). Several specific deletions of 1p have been identified including losses of 1p32.3, 1p31.3, 1p22.1–1p21.3 and 1p12 [9]. Clinically losses of 1p are a subset of secondary aberrations associated with a poor prognosis [9, 44, 51, 52]. The deletions of 1p21 (FAM46C) and 1p32 (CDKN2C) have been reported as independent prognostic factors [53, 54].

1q21 Amplification

At least three types of amplification of the 1q21 region are found in MM, all of which involve instability of the 1q12 pericentromeric heterochromatin. The first to be described was the intra-chromosomal amplification of 1q21 CN by direct and/or inverted duplications of the 1q12~23. A second mechanism for a much larger scale focal amplification of multiple copies of 1q21 in the proximal 1q12~23 region involves an inverted repeat pattern of amplification resulting from breakage-fusion-bridge (BFB) cycles mediated by the 1q12 pericentromeric heterochromatin [55]. The BFB cycle mechanism amplifies a large number of genes within an amplicon of about 10–15 Mb which spans the 1q12~23 region [56]. The bracketing of the amplicon by 1q12 pericentromeric heterochromatin allows the subsequent cycles of amplification to continue and high CNs of 1q21 to accumulate (Fig. 4.2d). CN gains of 1q12~q21 regions are most frequently found in proliferative, relapsed, and/or refractory myeloma in about 40% of cases [57]. Among the amplified and/or deregulated genes in this region are MCL1, MUC1, IL6R, BCL9, ANP32E, CKS1B, S100A4, PSMD4, and UBE2Q1, as well as many others [5, 6, 56–60]. The 1q12~23 amplicon is about 10 megabases in size and harbors one or more potential oncogenes. As a result of this type of amplification, interstitial copy number gains of 1q21 have been suggested as one of a number of potential "driver" lesions of myeloma progression due to the numerous potentially relevant oncogenes in this region [2].

A third mechanism for 1q21 amplification is caused by adding whole-arm segments of 1q to non-homologous chromosomes in what is known as jumping translocations of 1q (JT1q12) [43, 55, 61–64]. In patients with JT1q12s, it is known that the extra copies of the 1q21 result from the formation of triradial chromosomes involving the 1q12 region [43]. A triradial chromosome 1q12 is composed of a single short arm of chromosome 1p and two copies of 1q joined by 1q12 pericentromeric heterochromatin. Triradial chromosomes are transient intermediates which generate extra copies of the 1q, providing the donor chromosome 1q (DC).

The DC 1q segment breaks in the 1q12 region and can translocate (jump) to different receptor chromosomes (RCs) [43]. The JT1q12 translocations all result in an increase in the CN of 1q21 and all genes on the 1q.

Two types of JT1q12s are most common: the first occurs when the 1q jumps to the telomere of a RC, the second when the 1q12 jumps to the pericentromeric region of a RC, causing a whole-arm unbalanced translocation. Both of these jumping 1q12 aberrations cause partial trisomies for the 1q; however, the whole-arm translocations of 1q12 also cause large deletions in the RC, unlike the telomeric translocations. The most common whole-arm unbalanced translocations of 1q include the der(1;16)(q10;p10) (Fig. 4.2e), resulting in the amplification of 1q and loss of all of the 16q, and the der(1;19)(q10;p10), resulting in amp of 1q and the loss of 19q (Fig. 4.2f). Importantly, several recurring deletions reported in myeloma, such as deletions of 6q, 8p, and 17p, have all been reported to be associated with unbalanced whole-arm JT1q12s. The loss of 16q is associated with the LOH for *CYLD* and *WWOX* and whole-arm deletions which co-segregate in subclones with the gain of 1q. Clinically, the loss of 16q is associated with a worse overall survival [35]. These aberrations occur late in the progression of the disease, commonly in subclones, and show marked karyotypic instability and heterogeneity among cells within a tumor. These secondary unbalanced whole-arm aberrations of 1q12~23 may be favored during tumor progression, providing a selective advantage to subclones containing them.

The Origin of 1q21 Copy Number Increases

Genome-wide profiling studies provide evidence that a number of different epigenetic chromatin modifications occur during cancer progression [61]. In fact, simultaneous epigenetic influences include changes in DNA methylation, histone modifications, and nucleosome remodeling, as well as changes in small non-coding regulatory RNAs. It is known that DNA hypermethylation is a mechanism for the down regulation of genes, while DNA hypomethylation is associated with up regulation of gene expression. In MM, a gradual progression from overall global DNA hypomethylation to DNA hypermethylation of specific genes has been found [65–67]. In relation to the chromosomal aberrations the finding of global hypomethylation and especially the hypomethylation of highly repetitive elements is thought to contribute to tumor progression and genomic instability [68]. It has been hypothesized that hypomethylation of the 1q12 pericentromeric regions of metaphase chromosomes can induce JT1q12s. The 1q12 pericentromeric heterochromatin is distinctive in the human genome in that it is the largest single block of late-replicating, highly repetitive satII/III DNA, and is known to contain unstable segmental duplications (low copy repeats). Evidence for an epigenetic origin for high-risk 1q21 CN aberrations induced by the hypomethylation of the 1q12 pericentromeric region has been reported [69].

Clonal Evolution to Cytogenetically Defined High-Risk Disease

Genomic studies provide evidence that clonal evolution of MM involves both linear and non-linear patterns, and that minor subclones can survive treatment, acquire new anomalies, and provide a reservoir for relapse [2, 70, 71]. The use of early suboptimal treatment may eradicate the less aggressive drug-sensitive clones (hyperdiploid), while allowing the more aggressive drug-resistant clones with t(4;14), 17p− and +1q21 to expand [38]. These clones may survive because competition between the subclones for existing niches in the bone marrow microenviroment probably favors the genetic subtypes which contain secondary aberrations. In these clones, accumulation of aberrations is thought to occur in a Darwinian fashion and provide a selective pressure for the expansion of therapy resistant variants [2]. In fact, molecular profiling studies have shown that cytogenetically defined high-risk patients show more changes involving both intra-clonal and sub-clonal heterogeneity as well as alternating dominance of clones [70–73]. To help optimize clinical diagnostic testing and provide guidelines for cytogenetic risk categories, the International Myeloma Working Group (IMWG) has recommended the use of 3 markers, including the t(4;14), deletion 17p, and gain of 1q21 to identify high-risk patients [74]. Two of these markers, the t(4;14) and del17p, when identified together by interphase fluorescence in situ hybridization (iFISH), are considered cytogenetically defined high-risk disease. Metaphase FISH and SKY studies have shown that the co-segregation of the deletion of 17p and gain of 1q21 occur as secondary aberrations during the progression of clones with high-risk IGH translocations [72]. However these same high-risk secondary lesions also accumulate in "low-risk" progenitor clones with hyperdiploidy and no IGH translocation (Fig. 4.3a–d), demonstrating the need for continued cytogenetic surveillance during the disease. The clinical impact of these adverse lesions is presumably further modulated by the etiological background of the myeloma clone in which they occur, and therefore a +1q21 that is found in association with t(4;14) and del17p is thought to impart a worse prognosis than a +1q21 in association with hyperdiploidy. To date, the JT1q12 is the only known cytogenetic lesion with the potential to act as a "subclonal driver" resulting in sequential CNAs of itself (1q21), and potentially in multiple RCs as well [64].

Myelodysplastic Clones

Myelodysplastic syndrome (MDS) is a well-recognized complication of cancer chemotherapy. Cytogenetic aberrations which typify MDS following alkylator-based therapy include partial or complete deletions of 5, 7, 13, and 20, as well as +8, whole arm t(1;7)(q10;p10), and dic(5;17)(q11;p11). In MM, independent MDS clones are found following high dose therapy and show distinct differences from the complexity and ploidy levels found in MM (Table 4.3).

Table 4.3 Myelodysplastic clones

Deletions	Frequency (%)
del(20q)	10
−7/7q−	5
der(1;7)(q10;p10)	3
del(13q)/−13	3
+8	2
del(11q)	1
der(1;15)(q10;q10)	1
dic(5;17)(q11;p11)	1

Independent MDS clones usually show only a single abnormality, but in rare cases can show two or even three MDS aberrations [75]. The most common recurring aberrations which occur as solitary independent clones in addition to a MM clone include deletion 20q, monosomy and deletion of 13q, and der(1;7)(q10; p10). By far the most frequent solitary MDS clone found in MM is deletion 20q; it is reported in separate clones in 10% of karyotypically abnormal MM/MGUS patients [75].

Myelodysplastic chromosome aberrations also occur within complex myeloma karyotypes as secondary aberrations. These aberrations include partial or complete deletions of 5, 7, and 20, as well as +8 [76–78]. Of the MM clones showing MDS type aberrations, the most common aberrations were del(20q) (25%), −7/7q− (16%), t(1;7)(q10;p10) (13%), −5/5q− (10%), del(13q) (7%), +8 (5%), and del(11q) (2%) [79].

Conclusion

Considerable progress has been made in the genetic analysis of MM and it is now possible to identify cytogenetic and molecular subtypes of the disease. The continuing question is how to improve the current risk stratifications and generate better stratifications based on the use of improved iFISH probe combinations and molecular diagnostic tests. Currently the best cytogenetic assessment of risk stratification is achieved by performing both iFISH and conventional chromosome studies. It seems clear that chromosomal aberrations play a key role in clonal diversity of PC disorders and that cytogenetically more evolved clones with amplification of 1q21 are at a proliferative advantage.

Acknowledgments The interphase FISH images in Fig. 4.1 were kindly provided by Dr. Erming Tian, Myeloma Institute, University of Arkansas for Medical Sciences.

References

1. Kuehl WM, Bergsagel PL. Multiple myeloma: evolving genetic events and host interactions. Nat Rev Cancer. 2002;2(3):175–87.
2. Morgan GJ, Walker BA, Davies FE. The genetic architecture of multiple myeloma. Nat Rev Cancer. 2012;12(5):335–48.
3. Bergsagel PL, Kuehl WM, Zhan F, Sawyer J, Barlogie B, Shaughnessy J. Cyclin D dysregulation: an early and unifying pathogenic event in multiple myeloma. Blood. 2005;106 (1):296–303.
4. Zhan F, Huang Y, Colla S, Stewart JP, Hanamura I, Gupta S, Epstein J, Yaccoby S, Sawyer J, Burington B, Anaissie E, Hollmik K, Pineda-Roman M, Tricot G, van Rhee F, Walker R, Zangari M, Crowley J, Barlogie B, Shaughnessy JD Jr. The molecular classification of multiple myeloma. Blood. 2006;108(6):2020–2028.
5. Walker BA, Leone PE, Jenner MW, Li C, Gonzalez D, Johnson DC, Ross FM, Davies FE, Morgan GJ. Integration of global SNP-based mapping and expression arrays reveals key regions, mechanisms, and genes important in the pathogenesis of multiple myeloma. Blood. 2006;108(5):1733–1743.
6. Shaughnessy JD Jr, Zhan F, Burington BE, Huang Y, Colla S, Hanamura I, Stewart JP, Kordsmeier B, Randolph C, Williams DR, Xiao Y, Xu H, Epstein J, Anaissie E, Krishna SG, Cottler-Fox M, Hollmig K, Mohiuddin A, Pineda-Roman M, Tricot G, van Rhee F, Sawyer J, Alsayed Y, Walker R, Zangari M, Crowley J, Barlogie B. A validated gene expression model of high-risk multiple myeloma is defined by deregulated expression of genes mapping to chromosome 1. Blood. 2007;109(6):2276–84.
7. Avet-Loiseau H, Attal M, Moreau P, Charbonnel C, Garban F, Hulin C, Leyvraz S, Michallet M, Yakoub-Agha I, Garderet L, Marit G, Michaux L, Voillat L, Renaud M, Grosbois B, Guillerm G, Benboubker L, Monconduit M, Thieblemont C, Casassus P, Caillot D, Stoppa AM, Sotto JJ, Wetterwald M, Dumontet C, Fuzibet JG, Azais I, Dorvaux V, Zandecki M, Bataille R, Minvielle S, Harousseau JL, Facon T, Mathiot C. Genetic abnormalities and survival in multiple myeloma: the experience of the Intergroupe Francophone du Myelome. Blood. 2007;109(8):3489–96.
8. Avet-Loiseau H, Li C, Magrangeas F, Gouraud W, Charbonnel C, Harousseau J-L, Attal M, Marit G, Mathiot C, Facon T, Moreau P, Anderson KC, Campion L Munshi NC, Minvielle S. Prognostic significance of Copy-Number Alteration in multiple myeloma. J Clin Oncol. 2009;27(27):4585–4590.
9. Walker BA, Leone PE, Chiecchio L, Dickens NJ, Jenner MW, Boyd KD, Johnson DC, Gonzalez D, Paolo Dagrada G, Protheroe RKM, Konn ZJ, Stockley DM, Gregory WM, Davies FE, Ross FM, Morgan GJ. A compendium of myeloma associated chromosomal copy number abnormalities and their prognostic value. Blood. 2010;166(15):e56–65.
10. DeWald GW, Kyle RA, Hicks GA, Griepp PR. The clinical significance of cytogenetic studies in 100 patients with multiple myeloma, plasma cell leukemia or amyloidosis. Blood. 1985;66 (2):380–90.
11. Sawyer JR, Waldron JA, Jagannath S, Barlogie B. Cytogenetic findings in 200 patients with multiple myeloma. Cancer Genet Cytogenet. 1995;82(1):41–9.
12. Rajkumar SV, Greipp PR. Prognostic factors in multiple myeloma. Hematol Oncol Clin North Am. 1999;13(6):1295–314.
13. Zhan F, Sawyer J, Tricot G. The role of cytogenetics in multiple myeloma. Leukemia. 2006;20 (9):1484–1486.
14. Dewald GW, Therneau T, Larson D, Lee YK, Fink S, Smoley S, Paternoster S, Adeyinka A, Kettering R, Van Dyke DL, Fonseca R, Kyle R. Relationship of patient survival and chromosome anomalies detected in metaphase and/or interphase cells at diagnosis of myeloma. Blood. 2005;106(10):3553–3558.
15. Kapoor P, Fonseca R, Rajlkumar V, Sinha S, Gertz MA, Stewart K, Bergsagel L, Lacy MQ, Dingli DD, Ketterling RP, Buadi F, Kyle RA, Witzig TE, Greipp PR, Dispenzieri A, Kumar S.

Evidence for cytogenetic and fluorescence in situ hybridization risk stratification of newly diagnosed multiple myeloma in the era of novel therapies. Mayo Clin Poc. 2013;85(6):532–537.

16. Sawyer JR. The prognostic significance of cytogenetics and molecular profiling in multiple myeloma. Cancer Genet. 2011;204(1):3–12.

17. Nilsson T, Hoglund M, Lenhoff S, Rylander L, Turesson I, Westin J, Mitelman F, Johansson B. A pooled analysis of karyotypic patterns, breakpoints and imbalances in 783 cytogenetically abnormal multiple myelomas reveals frequently involved segments as well as significant age-and sex-related differences. Br J Haematol. 2003;120(6):960–9.

18. Chng WJ, Ketterling RP, Fonseca R. Analysis of genetic abnormalities provides insights into genetic evolution of hyperdiploid myeloma. Genes Chrom Cancer. 2006;45(12):1111–20.

19. Smadja NV, Fruchart C, Isnard F, Louvet C, Dutel JL, Cheron N, Grange MJ, Monconduit M, Bastard C. Chromosomal analysis in multiple myeloma: cytogenetic evidence of two different diseases. Leukemia. 1998;12(6):960–969.

20. Smadja NV, Bastard C, Brigaudeau C, Leroux D, Fruchart C. Hypodiploidy is a major prognostic factor in multiple myeloma. Blood. 2001;98(7):2229–2238.

21. Fonseca R, Barlogie B, Bataille R, Bastard C, Bergsagel PL, Chesti M, Davies FE, Drach J, Greipp PR, Kirsch IR, Kuehl WM, Hernandez JM, Minvielle S, Pilarski LM, Shaughnessy JD, Stewart AK, Avet-Loiseau H. Genetics and cytogenetics of multiple myeloma: a workshop report. Cancer Res. 2004;64(4):1546–58.

22. Drach J, Schuster J, Nowotny H, Angerler J, Rosenthal F, Fiegl M, Rothermundt C, Gsur A, Jager U, Heinz R. Multiple myeloma: high incidence of chromosomal aneuploidy as detected by interphase fluorescence in situ hybridization. Cancer Res. 1995;55(17):3854–3859.

23. Tonon G. Moleular pathogenesis of multiple myeloma. Hematol Oncol Clin North Am. 2007;21(6):985–1006.

24. Bergsagel PL, Chesi M, NardinI E, Brents LA, Kirby SL, Kuehl WM. Promiscuous translocation into immunoglobulin heavy chain switch regions in multiple myeloma. Proc Natl Acad Sci USA. 1996;93(24):13931–6.

25. Bergsagel PL, Kuehl WM. Chromosomal translocations in multiple myeloma. Oncogene. 2001;20(40):5611–5622.

26. Chesi M, Bergsagel PL, Brents LA, Smith CM, Gerhard DS, Kuehl WM. Dysregulation of cyclin D1 by translocation into an IgH gamma switch region in two multiple myeloma cell lines. Blood. 1996;88(2):674–81.

27. Chesi M, Nardini E, Brents LA, Schröck E, Ried T, Kuehl WM, Bergsagel PL. Frequent translocation t(4;14)(p16.3;q32.3) in multiple myeloma is associated with increased expression and activating mutations of fibroblast growth factor receptor 3. Nat Genet. 1997;3:260–4.

28. Tian E, Sawyer JR, Heuck CJ, Zhang Q, van Rhee F, Barlogie B, Epstein J. In multiple myeloma, 14q32 translocations are nonrandom chromosomal fusions driving high expression levels of the respective partner genes. Genes Chrom Cancer. 2014;53(7):549–57.

29. Bergsagel PL, Kuehl WM. Molecular pathogenesis and a consequent classification of multiple myeloma. J Clin Oncol. 2005;23(26):6333–6338.

30. Keats JJ, Reiman T, Maxwell CA, Taylor BJ, Larratt LM, Mant MJ, Belch AR, Pilarski LM. In multiple myeloma t(4;14)(p16;q32) is an adverse prognostic factor irrespective of FGRR3 expression. Blood. 2003;101(4):1520–9.

31. Fonseca R, Blood E, Rue M, Harrington D, Oken MM, Kyle RA, Dewald GW, Van Ness B, Van Wier SA, Henderson KJ, Bailey RJ, Greipp PR. Clinical and biologic implications of recurrent genomic aberrations in myeloma. Blood. 2003;101(11):4569–73.

32. Gertz MA, Lacy MQ, Dispenzieri A, Greipp PR, Litzow MR, Henderson KJ, Van Wier SA, Ahmann GJ, Fosseca R. Clinical implications of t(11;14)(q13;q32), t(4;14)(p16.3;q32), and -17p13 in myeloma patients treated with high-dose therapy. Blood. 2005;106(8):2837–41.

33. Avet-Loiseau H, Facon T, Daviet A, Godon C, Rapp MJ, Harousseau JL, Grosbois B, Bataille R. 14q32 translocations and monosomy 13 observed in monoclonal gammopathy of undertermined significance delineate a multistep process for the oncogensis of multiple myeloma. Cancer Res. 1999;59(18):4546–4550.

34. Chesi M, Bergsagel PL, Shonukan OO, Martelli ML, Brents LA, Chen T, Schröck E, Ried T, Kuehl WM. Frequent dysregulation of the c-maf proto-oncogene at 16q23 by translocation to an Ig locus in multiple myeloma. Blood. 1998;91(12):4457–63.
35. Jenner MW, Leone PE, Walker BA, Ross FM, Johnson DC, Gonzalez D, Chiecchio L, Dachs Cabanas E, Dagrada GP, Nightingale M, Protheroe RK, Stockley D, Else M, Dickens NJ, Cross NC, Davies FE, Morgan GJ. Gene mapping and expression analysis of 16q loss of heterozygosity identifies WWOX and CYLD as being important in determining clinical outcome in multiple myeloma. Blood. 2007;110(9):3291–30.
36. Rajkumar SV, Gupta V, Fonseca R, Dispenzieri A, Gonsalves WI, Larson D, Ketterling RJ, Lust JA, Kyle RA, Kumar SK. Impact of primary molecular cytogenetic abnormalities and risk of progression in smoldering multiple myeloma. Leuk. 2013;27(8):1738–44.
37. Kaufmann H, Ackermann J, Baldia C, Nösslinger T, Wieser R, Seidl S, Sagaster V, Gisslinger H, Jäger U, Pfeilstöcker M, Zielinski C, Drach J. Both IGH translocations and chromosome 13q deletions are early events in monoclonal gammopathy of undetermined significance and do not evolve during transition to multiple myeloma. Leukemia. 2004;11:1879–82.
38. Bergsagel PL, Chesi MV. Molecular classification and risk stratification of myeloma. Hematol Oncol. 2013;3(Suppl 1):38–41.
39. López-Corral L, Gutiérrez NC, Vidriales MB, Mateos MV, Rasillo A, García-Sanz R, Paiva B. San Miguel JF. The progression from MGUS to smoldering myeloma and eventually to multiple myeloma involves a clonal expansion of genetically abnormal plasma cells. Clin Cancer Res. 2011;17(7):1692–700.
40. Gabrea A, Martelli ML, Qi Y, Roschke A, Barlogie B, Shaughnessy JD Jr, Sawyer JR, Kuehl WM. Secondary genomic rearrangements involving immunoglobulin or MYC loci show similar prevalences in hyperdiploid and nonhyperdiploid myeloma tumors. Genes Chrom Cancer. 2008;47(7):573–590.
41. Walker BA, Wardell CP, Brioli A, Boyle E, Kaiser MF, Begum DB, Dahir NB, Johnson DC, Ross FM, Davies FE, Morgan GJ. Translocations at 8q24 juxtapose MYC with genes that harbor superenhancers resulting in overexpression and poor prognosis in myeloma patients. Blood Cancer J. 2014;4:e191.
42. Affer M, Chesi M, Chen WD, Keats JJ, Demchenko YN, Tamizhmani K, Garbitt VM, Riggs DL, Brents LA, Roschke AV, Van Wier S, Fonseca R, Bergsagel PL, Kuehl WM. Promiscuous MYC locus rearrangements hijack enhancers but mostly super-enhancers to dysregulate MYC expression in multiple myeloma. Leukemia. 2014;28(8):1725–35.
43. Sawyer JR, Tricot G, Mattox S, Jagannath S, Barlogie B. Jumping translocations of chromosome 1q in multiple myeloma: evidence for a mechanism involving decondensation of pericentromeric heterochromatin. Blood. 1998;91(5):1732–41.
44. Wu KL, Beverloo B, Lokhorst HM, Segeren CM, van der Holt B, Steijaert MM, Westveer PH, Poddighe PJ, Verhoef GE, Sonneveld P. Dutch-Belgian Haemato-Oncology Cooperative Study Group (HOVON); Dutch Working Party on Cancer Genetics and Cytogenetics (NWCGC). Abnormalities of chromosome 1p/q are highly associated with chromosome 13/13q deletions and are an adverse prognostic factor of the outcome of high-dose chemotherapy in patients with multiple myeloma. Br J Haematol. 2007;136(4):615–623.
45. Tricot G, Barlogie B, Jagannath S, Bracy D, Mattox S, Vesole DH, Naucke S, Sawyer JR. Poor prognosis in multiple myeloma is associated only with partial or complete deletions of chromosome 13 or abnormalities involving 11q and not the other karyotype abnormalities. Blood. 1995;86(11):4250–6.
46. Tricto G, Sawyer JR, Jagannath S, Desikan KR, Siegel D, Naucke S, Mattox S, Bracy D, Munshi N, Barlogie B. Unique role of cytogenetics in the prognosis of patients with myeloma receiving high-dose therapy and autotransplants. J Clin Oncol. 1997;15(7):2659–2666.
47. Perez-Simon JA, Garcia-Sanz R, Tabernero MD, Almeida J, Gonzalez M, Fernandez-Calvo J, Moro Mj, Hernandez JM, San Miguel JF, Orfao A. Prognostic value of numerical chromosome aberrations in multiple myeloma: a FISH analysis of 15 different chromosomes. Blood. 1998;91(9):3366–3371.

48. Shaugnessy JD, Tian E, Sawyer J, Bumm K, Landes R, Badros A, Morris C, Tricot G, Epstein J, Barlogie B. High incidence of chromosome 13 deletion in multiple myeloma detected by multiprobe interphase FISH. Blood. 2000;96(4):1505–11.
49. Drach J, Ackermann J, Fritz E, Kromer E, Schuster R, Gisslinger H et al. Presence of a p53 gene deletion in patients with multiple myeloma predicts for short survival after conventional–dose chemotherapy. Blood. 1998;92(1):803–809.
50. Cremer FW, Bila J, Buck I, Kartal M, Hose D, Ittrich C, Benner A, Raab MS, Theil A-C, Moos M, Goldschmidt H, Bartram CR, Jauch A. Delineation of distinct subgroups of multiple myeloma and a model for clonal evolution based on interphase cytogenetics. Genes Chrom Cancer. 2005;44(2):194–203.
51. Debes-Marun CS, Dewald GW, Bryant S, Picken E, Santana-Davila R, González-Paz N, Winkler JM, Kyle RA, Gertz MA, Witzig TE, Dispenzieri A, Lacy MQ, Rajkumar SV, Lust JA, Greipp PR, Fonseca R. Chromosome abnormalities clustering and its implication for pathogenesis and prognosis in myeloma. Leukemia. 2003;17(2):427–26.
52. Chang H, Ning Y, Qi X, Yeung J, Xu W. Chromosome 1p21 deletion is a novel prognostic marker in patients with multiple myeloma. Br J Haematol. 2007;(1):139:51–54.
53. Boyd KD, Ross FM, Walker BA, Wardell CP, Tapper WJ, Chiecchio L, Dagrada G, Konn ZJ, Gregory WM, Jackson GH, Child JA, Davies FE, Morgan GJ. Mapping of chromosome 1p deletions in myeloma identifies FAM46C at 1p12 and CDKN2C at 1p32.3 as being genes in regions associated with adverse survival. Clin Cancer Res. 2011;17(24):7776–84.
54. Hebraud B, Leleu X, Lauwers-Cances V, Roussel M, Caillot D, Marit G, Karlin L, Hulin C, Gentil C, Guilhot F, Garderet L, Lamy T, Brechignac S, Pegourie B, Jaubert J, Dib M, Stoppa AM, Sebban C, Fohrer C, Fontan J, Fruchart C, Macro M, Orsini-Piocelle F, Lepeu G, Sohn C, Corre J, Facon T, Moreau P, Attal M, Avet-Loiseau H. Deletion of the 1p32 region is a major independent prognostic factor in young patients with myeloma: the IFM experience on 1195 patients. Leukemia. 2014;3:675–9.
55. Sawyer JR, Tian E, Thomas E, Koller M, Stangeby C, Sammartino G, Gossen L, Swanson CM, Binz RL, Barlogie B, Saughnessy J. Evidence for a novel mechanism for gene amplification in multiple myeloma: 1q12 pericentromeric heterochromatin mediates breakage-fusion-bridge cycles of a 1q12 ~ 23 amplicon. Br. J Haematol. 2009;147(4):484–94.
56. Carrasco DR, Tonon G, Huang Y, Zhang Y, Sinha R, Feng B, Stewart JP, Zhan F, Khatry D, Protopopova M, Protopopov A, Sukhdeo K, Hanamura I, Stephens O, Barlogie B, Anderson KC, Chin L, Shaughnessy JD Jr, Brennan C, Depinho RA. High-resolution genomic profiles defines distinct clinico-pathogenetic subgroups of multiple myeloma patients. Cancer Cell. 2006;9(4):313–325.
57. Hanamura I, Stewart JP, Huang Y, Zhan F, Santra M, Sawyer JR, Hollmig K, Zangarri M, Pineda-Roman M, van Rhee F, Cavallo F, Burington B, Crowley J, Tricot G, Barlogie B, Shaughnessy JD Jr. Frequent gain of chromosome band 1q21 in plasma cell dyscrasias detected by fluorescence in situ hybridization: incidence increases from MGUS to relapsed myeloma and is related to prognosis and disease progression following tandem stem cell transplantations. Blood. 2006;108(5):1724–32.
58. Treon SP, Maimonis P, Bua D, Young G, Raje N, Mollick J, Chauhan D, Tai YT, Hideshima T, Shima Y, Hilgers J, von Mensdorff-Pouilly S, Belch AR, Pilarski LM, Anderson KC. Elevated soluble MUC1 levels and decreased anti-MUC1 antibody levels in patients with multiple myeloma. Blood. 2000;96(9):3147–53.
59. Zhang B, Gojo I, Fenton RG. (2002) Myeloid cell factor-1 is a critical survival factor for multiple myeloma. Blood. 2002;99(6):1885–1893.
60. Inoue J, Otsuki T, Hirasawa A, Imoto I, Matsuo Y, Shimizu S, Taniwaki M, Inazawa J. Overexpression of PDZK1 within the 1q12-q22 amplicon is likely to be associated with drug-resistance phenotype in multiple myeloma. Am J Path. 2004;165(1):71–81.

61. Le Baccon P, Leroux D, Dascalescu C, Duley S, Marais D, Esmenjaud E, Sotto JJ, Callanan M. Novel evidence of a role for chromosome 1 pericentric heterochromatin in the pathogenesis of B-cell lymphoma and multiple myeloma. Genes Chrom Cancer. 2001;32 (3):250–64.

62. Sawyer JR, Tricot G, Lukacs JL, Binz RL, Tian E, Barlogie B, Shaughnessy J Jr. Genomic instability in multiple myeloma: evidence for jumping segmental duplications of chromosome arm 1q. Genes Chrom Cancer. 2005;42(1):95–106.

63. Fournier A, Florin A, Lefebvre C, Solly F, Leroux D, Callanan MB. Genetics and epigenetics of 1q rearrangments in hematological malignancies. Cytogenet Genome Res. 2007;118(2–4):320–7.

64. Sawyer JR, Tian E, Heuck CJ, Epstein J, Johann DJ, Swanson CM, Lukacs JL, Johnson M, Litchi-Binz R, Boast A, Sammartino G, Usmani S, Zangari M, Waheed S, van Rhee F, Barlogie B. Jumping translocations of 1q12 in multiple myeloma: a novel mechanism for deletion of 17p in cytogenetically defined high-risk disease. Blood. 2014;123(16):2504–12.

65. You JS, Jones PA. Cancer genetics and epigenetics: two sides of the same coin? Cancer Cell. 2012;22(1):9–20.

66. Walker BA, Wardell CP, Chiecchio L, Smith EM, Boyd KD, Neri A, Davies FE, Ross FM, Morgan GJ. Aberrant global methylation patterns affect the molecular pathogenesis and prognosis of multiple myeloma. Blood. 2011;117(2):553–62.

67. Heuck CJ, Mehta J, Bhagat T, Gundabolu K, Yu Y, Khan S, Chrysofakis G, Schinke C, Tariman J, Vickrey E, Pulliam N, Nischal S, Zhou L, Bhattacharyya S, Meagher R, Hu C, Maqbool S, Suzuki M, Parekh S, Reu F, Steidl U, Greally J, Verma A, Singhal SB. Myeloma is characterized by stage-specific alterations in DNA methylation that occur early during myelomagenesis. J Immunol. 2013;190(6):2966–75.

68. Bollati V, Fabris S, Pegoraro V, Ronchetti D, Mosca L, Deliliers GL, Motta V, Bertazzi PA, Baccarelli A, Neri A. Differential repetitive DNA methylation in multiple myeloma molecular subgroups. Carcino. 2009;8:1330–5.

69. Sawyer JR, Tian E, Heuck CJ, Johann DJ, Epstein J, Swanson CM, Lukacs JL, Binz RL, Johnson M, Sammartino G, Zangari M, Davies FE, van Rhee F, Morgan GJ, Barlogie B. Evidence of an epigenetic origin for high-risk 1q21 copy number aberrations in multiple myeloma. Blood. 2015;125(24):3756–9.

70. Keats JJ, Chesi M, Egan JB, Garbitt VM, Palmer SE, Braggio E, Van Wier S, Blackburn PR, Baker AS, Dispenzieri A, Kumar S, Rajkumar SV, Carpten JD, Barrett M, Fonseca R, Stewart AK, Bergsagel PL. Clonal competition with alternating dominance in multilple myeloma. Blood. 2012;120(5):1067–76.

71. Magrangeas F, Avet-Loiseau H, Gouraud W, Lodé L, Decaux O, Godmer P, Garderet L, Voillat L, Facon T, Stoppa AM, Marit G, Hulin C, Casassus P, Tiab M, Voog E, Randriamalala E, Anderson KC, Moreau P, Munshi NC, Minvielle S. Minor clone provides a reservoir for relapse in multiple myeloma. Leuk. 2013;27(2):473–81.

72. Boyd KD, Ross FM, Chiecchio L, Dagrada GP, Konn ZJ, Tapper WJ, Walker BA, Wardell CP, Gregory WM, Szubert AJ, Bell SE, Child JA, Jackson GH, Davies FE, Morgan GJ. A novel prognostic model in myeloma based on co-segregating adverse FISH lesions and the ISS: analysis of patients treated in the MRC Myeloma IX trial. Leukemia. 2012;2:349–55.

73. Pawlyn C, Melchor L, Murison A, Wardell CP, Brioli A, Boyle EM, Kaiser MF, Walker BA, Begum DB, Dahir NB, Proszek P, Gregory WM, Drayson MT, Jackson GH, Ross FM, Davies FE, Morgan GJ. Coexistent hyperdiploidy does not abrogate poor prognosis in myeloma with adverse cytogenetics and may precede IGH translocations. Blood. 2015;125 (5):831–40.

74. Chng WJ, Dispenzieri A, Chim C-S, Fonseca R, Goldschmidt H, Lentzch S, Munshi N, Palumbo A, Miguel JS, Sonneveld P, Cavo M, Usmani S, Durie BGM, Avet-Loiseau H. IMWG consensus on risk stratification in multiple myeloma. Leukemia. 2014;28(2):269–77.

75. Nilsson T, Nilsson L, Lenhoff S, Rylander L, Astrand-Grundstrom I, Strombeck B, Hoglund M, Turesson I, Westin J, Mitelman F, Jacobsen SEW, Johansson B. MDS/AML-associated cytogenetic abnormalities in multiple myeloma and monoclonal gammopathy of undetermined significance: Evidence for frequent de novo occurrence and multipotent stem cell involvement of del(20q). Genes Chrom Cancer. 2004;41(3):223–31.
76. Govindarajan R, Jagannath S, Flick JT, Vesole DH, Sawyer J, Barlogie B, Tricot G. Preceding standard therapy is the likely cause of MDS after autotransplants for multiple myeloma. Br J Haematol. 1996;95(2):349–53.
77. Jacobson J, Barlogie B, Shaughnessy J, Drach J, Tricot G, Fassas A, Zangari M, Giroux D, Crowley J, Hough A, Sawyer J. MDS-type abnormalities with myeloma signature karyotype (MM-MDS): only 13% 1-year survival despite tandem transplants. Br J Haematol. 2003;122 (3):430–40.
78. Usmani SZ, Sawyer J, Rosenthal A, Cottler-Fox M, Epstein J, Yaccoby S, Sexton R, Hoering A, Singh Z, Heuck CJ, Waheed S, Chauhan N, Johann D, Abdallah AO, Muzaffar J, Petty N, Bailey C, Crowley J, van Rhee F, Barlogie B. Risk factors for MDS and acute leukemia following total therapy 2 and 3 for multiple myeloma. Blood. 2013;121(23):4753–7.
79. Barlogie B, Tricot G, Haessler J, van Rhee F, Cottler-Fox M, Anissie E, Waldron J, Pineda-Roma M, Thertulien R, Zangari M, Hollmig K, Mohiuddin A, Alsayed Y, Hoering A, Crowley J, Sawyer J. Cytogenetically defined myelodysplasia after melphalan-based autotransplantation for multiple myeloma linked to poor hematopoietic stem-cell mobilization: the Arkansas experience in more than 3000 patients treated since 1989. Blood. 2008;111(1):94–100.

Chapter 5
Flow Cytometric Analysis in the Diagnosis and Prognostication of Plasma Cell Neoplasms

Pei Lin

FCM for Diagnosis of PCN

Plasma cell neoplasms encompass a range of diseases as a result of proliferation by monoclonal plasma cells. According to the WHO classification, these include monoclonal gammopathy of undermined significance (MGUS), myeloma, solitary plasmacytoma of bone, extraosseous plasmacytoma and monoclonal immunoglobulin deposition diseases. Regardless of the disease categories, PCs in these neoplasms share similar immunophenotypic features and are distinct from those of normal PCs. FCM allows assessment of a large number of events for multiple antigens simultaneously, including surface and cytoplasmic light chains, to distinguish normal or reactive PCs from neoplastic PCs. Typically CD38, CD138, CD45, and light scatters are used to identify all PCs, benign or malignant. CD38 is brighter on PCs than other hematopoietic components. CD138 expression is normally restricted to PCs within the hematopoietic system. However, CD38 can be decreased or rarely absent, and CD138 is not expressed in all myeloma cases. The combination of CD38 and CD138 allows more reliable identification of PCs. CD138, also known as Syndecan-1 is a type I transmembrane heparan sulfate proteoglycan that normally undergoes proteolytic cleavage and is shed into the extracellular environment. Samples should be kept in room temperature and ideally processed within 24–48 h to preserve the antigen. PCs are usually dim or negative for CD45 and have a higher SSC than lymphocytes.

Compared to normal PCs, neoplastic PCs typically show the following immunophenotypic aberrancies: absent or decreased expression of CD27, CD45 and CD81; overexpression of CD20, CD28, CD33, CD56, CD117, CD200, and CD307. CD38 is usually decreased. They also express monoclonal cytoplasmic

P. Lin (✉)
Department of Hematopathology, University of Texas-MD Anderson Cancer Center, 1515 Holcombe Blvd, Houston, TX 77030, USA
e-mail: peilin@mdanderson.org

© Springer International Publishing Switzerland 2016
R.B. Lorsbach and M. Yared (eds.), *Plasma Cell Neoplasms*,
DOI 10.1007/978-3-319-42370-8_5

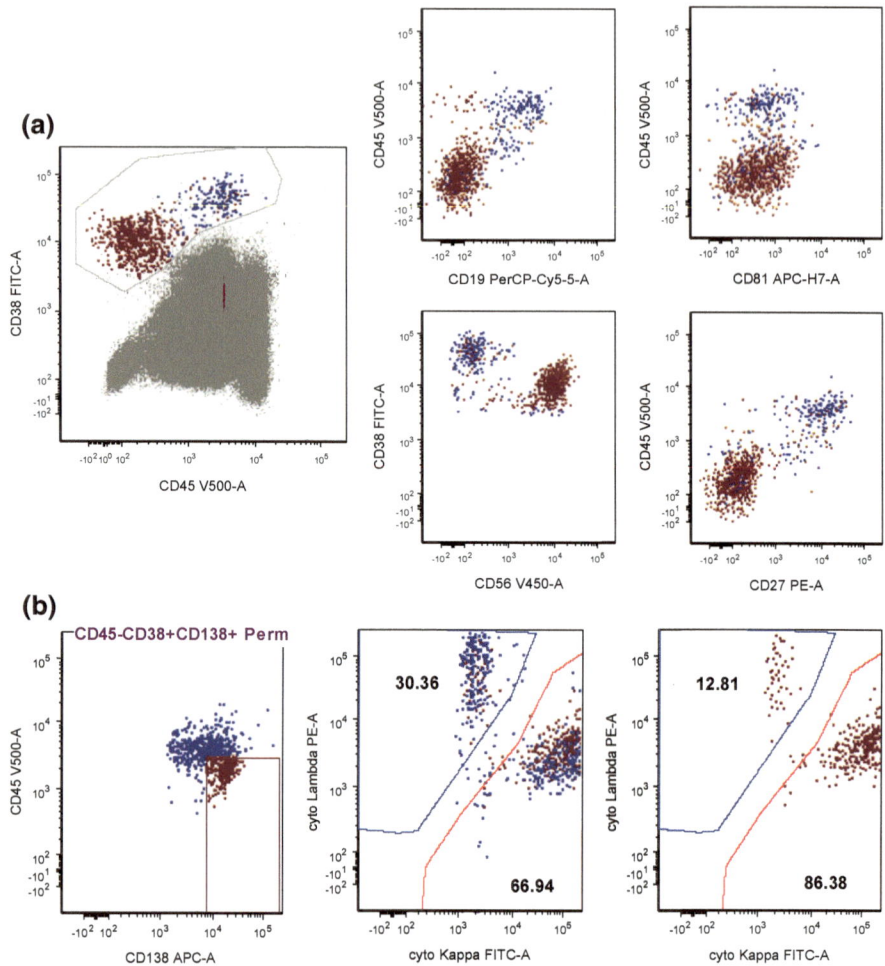

Fig. 5.1 a A case illustrates the immunophenotypic profiles of neoplastic plasma cells and reactive/benign plasma cells. The neoplastic plasma cells are CD38+ [diminished], CD138+, CD56 +, CD45−, CD19−, CD27−, CD81−. *Blue* Normal plasma cells, *Red* aberrant plasma cells. **b** Selective gating on CD45−/[dim] plasma cells allows identification of cytoplasmic light chain restricted monotypic PCs from a background of admixed polytypic PCs

immunoglobulin light chain [1] (Fig. 5.1a). Rarely, PCs may weakly express surface light chains. They are usually negative for B cell markers, such as CD19 or CD22. The neoplastic PCs may also show increased SSC and FSC.

Normal PCs from bone marrow of healthy donors, reactive, or from regenerating marrow of patients following therapy for non-PC related disease share an immunophenotype that is distinct from the vast majority of clonal PCs. They are usually positive for CD19, CD27, CD45, and CD81. Analysis of cytoplasmic immunoglobulin light chain further helps to distinguish neoplastic PCs from benign

PCs when combined with surface markers, particularly in setting of MGUS or post therapy where admixed polytypic PCs may comprise a significant fraction of the total PC population. In this regard, FCM is more sensitive than immunohisto-chemistry as gating specifically on CD19–CD45−/^{dim} PCs improves the sensitivity of cytoplasmic immunoglobulin light chain analysis (Fig. 5.1b).

A subset of normal PCs is CD19 negative or CD45dim/negative, phenotypically overlapping with that of neoplastic PCs. There may be a small subset positive for CD20, CD28, and CD56. A marker of activated T cells, CD28 can be overex-pressed in normal PCs in reactive conditions such as after chemotherapy [2]. One study found that CD19 was negative in 50% of normal individuals with a mean 24% normal PCs being negative. Most myelomatous PCs are negative for CD19, so for identification of aberrant PCs, it has the highest sensitivity of nearly 100% but only 50% specificity [3]. Rarely, myelomatous cells may express CD19. Similarly, CD45 has a high sensitivity of 90% but only 59% specificity. CD20 has a high specificity of 91% but only 34% sensitivity, while CD56 has a modest sensitivity of 69% and specificity of 74%. Still there are several distinguishing features that can help distinguish malignant PCs from their benign counterparts. Compared to neo-plastic PCs, expression of "aberrant markers" such as CD28 or CD56 is usually weak or heterogeneous spanning a spectrum of positive to negative on normal PCs

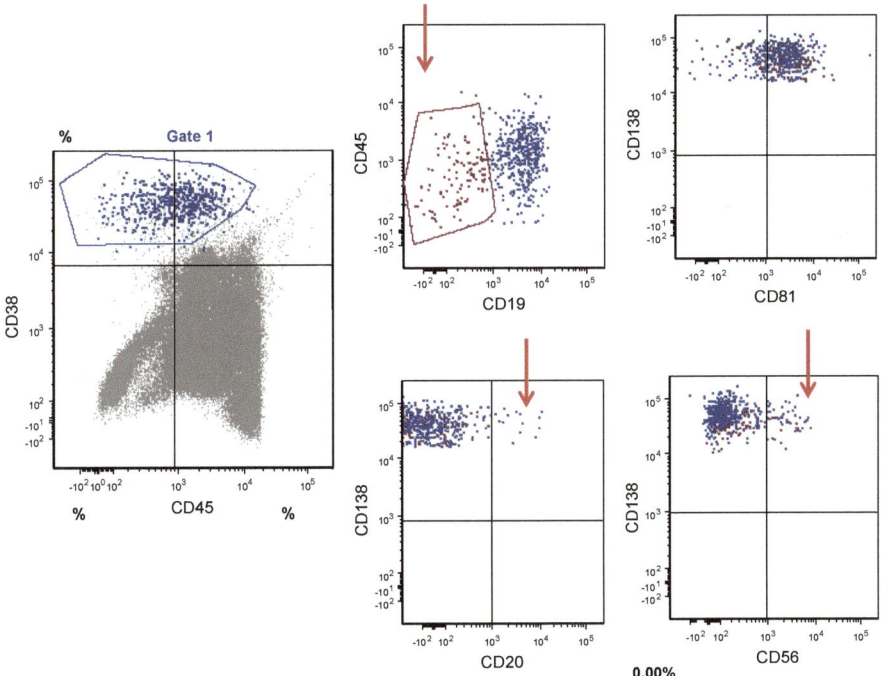

Fig. 5.2 A case illustrates normal PCs showing atypical immunophenotype in a small subset (CD56 weakly positive, CD19dim+, CD45 variable+)

(Fig. 5.2). By contrast, expression of these antigens is usually more uniform in malignant PCs [2, 3]. Furthermore, malignant PCs usually show more aberrancies within a defined population than 1–2 isolated "aberrancies" that spread among different subpopulations. For example, usually CD19 and CD20 are not coexpressed on normal PCs. CD19− or CD56+ normal PCs are usually not simultaneously CD27^{dim}/− or CD81^{dim}/−, allowing their distinction from myelomatous PCs. Thus, assessment of multiple antigens simultaneously is necessary to achieve a high sensitivity and specificity.

For clinical diagnosis, a basic panel commonly includes CD38, CD138, CD45, CD19, CD56, and cytoplasmic κ and λ immunoglobulin light chain. Addition of CD27, CD28, CD81, and CD117 further improves the discriminating power of FCM. Overall, CD19 expressing myeloma can be seen in <5% of cases of myeloma; while CD20+ and CD45+ myeloma represent about 20 and 10% of the cases, respectively. CD56 is typically bright in 60–75% of cases, or dim in about 10–15% of cases. CD117 can be detected in about 25–32%. Other antigens such as CD28 and CD33 are expressed in 36 and 18% of cases, respectively [4]. CD200 is expressed in 75–90% of cases. CD27 is aberrantly dim or lost in about 50% of myelomas cases. About 55% of myelomas show diminished CD81 [5–7].

FCM for Identification of Biomarkers

FCM is quantitative and allows estimation of the frequency of neoplastic PCs relative to benign PCs. Though FCM underestimates PCs due to various reasons, there is usually a correlation between the tumor load and the number of PCs identified by FCM. The fraction of normal PCs is usually >5% of total PCs in the majority of MGUS cases and yet fewer than 15% of symptomatic myeloma patients have such a finding at diagnosis. The fraction of neoplastic cells expands as MGUS or smoldering myeloma progresses to symptomatic myeloma [8, 9]. By quantifying the neoplastic fraction, FCM may help assess the risk of progression of low-grade indolent disease to symptomatic myeloma [10]. Currently, a ratio of >95% is considered critical. Cases of MGUS that have a high ratio of aberrant PCs relative to normal PCs despite a low PC count may have a higher risk of progression to symptomatic myeloma. Conversely, patients with symptomatic disease who have >5% of normal PCs tend to have less aggressive clinical features with higher hemoglobin levels, lower M protein levels and a lower frequency of high-risk cytogenetic abnormalities such as del(17). In addition, these patients are more likely to respond to therapy and achieve significantly longer progression free survival and overall survival. Calculation of this ratio is also useful in post treatment settings. In patients who have achieved complete remission (CR), immune reconstitution is usually accompanied by repopulation of normal polytypic PCs in the bone marrow, a sign of recovery from prior therapy.

About 8% of symptomatic myeloma patients remain stable and enjoy long-term survival of over 10 years with no clinically overt relapse, despite not having achieved a strict CR [11]. In such patients, the neoplastic PCs have an immunophenotypic signature resembling that of MGUS with a high ratio of normal to aberrant PCs upon principal component analysis [12]. The "MGUS signature" may help then to identify patients for whom more aggressive treatment to eradicate all evidence of MRD is not needed. By contrast, patients with persistent disease lacking such a signature are more likely to carry drug resistant clones requiring more aggressive intervention to prevent relapse.

The prognostic significance of different immunophenotypic subtypes is a subject of debate. However, immunophenotypic profiles may be related to the underlying genetic aberrations and DNA ploidy [4]. The non-hyperdiploid tumors are more likely negative for CD56 and CD117 but positive for CD20 and CD28. Tumors with 13q deletion and immunoglobulin heavy chain (IGH) gene rearrangements usually lack CD117. Tumors with t(11;14) (*CCND1-IGH*) fusion often express CD20 and lack CD56 and CD117. By contrast, tumors with t(4;14) or t"(4;16) usually lack CD20 [13]. Expression of CD117 and a lack of both CD19 and CD20 correlate with hyperdiploidy but without recurrent 14q32 IGH gene translocations. Tumors lacking CD19, CD20 and CD27 are associated frequently with t(4;14) or t(14;16) and non-hyperdiploidy [14].

CD19 or CD20 expression is associated both the t(11;14) and small cell morphology, in which the neoplastic cells have plasmacytoid cytologic features. PC leukemia also tends to express CD20 and have a small cell type of morphology.

Investigators have also shown that among patients treated with high dose chemotherapy and stem cell transplantation, expression of both CD19 and CD28 as well as the absence of CD117 were associated with a significantly shorter progression free survival and overall survival. CD28 expression correlated with non-hyperdiploid, t(14;16) and del(17p), while CD117-negative patients were associated with t(4;14) and del(13q). Patients can be stratified into three risk levels based on pattern of CD28 and CD117: poor risk (CD28 positive CD117 negative), intermediate (both negative and both positive), and good risk (CD28 negative CD117 positive) [4, 13, 14]. CD27 expression is reported to be associated with a better prognosis. The significance of CD45 or CD56 expression as a prognostic markers is controversial.

It should be noted that new therapies, such as proteasome inhibitors, immune modulating drugs (IMiDs), and antibody immunotherapy are dramatically changing the therapeutic options for myeloma patients, for whom the mainstay of therapy for many years was largely confined to high dose chemotherapy and stem cell transplant. Whether the prognostic significance of different immunophenotypic profiles described before the era of novel therapy still holds true remains to be investigated. The current risk stratification is primarily based on the presenting clinical features as well as cytogenetic and molecular characteristics.

Distinct immunophenotypic subclones can be detected in 30% of newly diagnosed myelomas, which may correspond to different subclones within a tumor. The identification of phenotypic subclones, therefore, may also reveal underlying genetic information that could guide therapy. Using a panel of antibodies including CXCR4, CD44, CD19, HLA-DR, CD54, CD49e, CD138, β7, CD33, CD20 and

CD81, investigators found that CD56, CD27, and CD81 are the most useful markers for studying intratumor heterogeneity of myeloma [15]. Interestingly, this study showed that clonogenic potential is not restricted to any specific immunophenotypic marker as both CD45+ and CD45− subsets can be clonogenic.

Circulating aberrant PCs have been detected at diagnosis in patients with PCNs [16]. However, whether levels of circulating PCs are directly related to tumor burden, stage of disease or underlying biology are still under investigation. Studies are also ongoing to investigate if aberrant PCs identified in stem cell products collected from the peripheral blood for patients undergoing stem cell transplant could affect the outcome.

FCM for MRD Monitoring

Stringent CR (sCR) was originally defined by the International Myeloma Working Group (IMWG) as absence of a serum M protein, normalization of the serum free light chain (sFLC) ratio, <5% of PCs in the bone marrow, and absence of monoclonal PCs in the BM as assessed by either immunohistochemistry or immunofluorescence [17]. However, MRD is still detectable in a subset of patients in sCR by more sensitive methods such as FCM or molecular studies. Similar to high-risk genetic features in myeloma, persistent MRD correlates with an inferior survival [18–20]. By contrast, absence of MRD, the so called immunophenotypic remission, correlates with a better survival in patients treated with high dose chemotherapy and stem cell transplantation regardless of serum immunofixation results. The benefits of immunophenotypic remission are evident in patients of both high-risk and standard-risk groups classified according to their cytogenetic profiles. Thus, the IMWG has updated the current definition of sCR to include absence of aberrant PCs by 4-color FCM.

MRD testing by FCM is based on the principle that neoplastic PCs can be reliably distinguished from normal PCs when assessed by multiple parameters, and in conjunction with cytoplasmic immunoglobulin light chain evaluation as needed. Technological advances in recent years have improved sensitivity of FCM by simultaneous analysis of eight or more markers and acquisition of over 2 million events. Consequently, FCM assessment of MRD can achieve a sensitivity level of 10^{-5} or higher. FCM is applicable in nearly all myeloma patients, even when the original tumor immunophenotype is unknown. It has a rapid turnaround time. Furthermore, using internal controls, such as B cell precursors, erythroblasts, myeloid precursors, and/or mast cells, sample quality can be assessed prior to MRD assessment to identify potential false-negative results.

The International Clinical Cytometry Society (ICCS) and European Society for Clinical Cell Analysis (ESCCA) have outlined consensus guidelines on specimen quality, staining process, reagent combinations, and the data acquisition process for MRD testing [1, 21, 22]. The consensus panel for MRD testing recommended includes CD138 and CD38, CD45, CD19, CD56, CD27, CD81, and CD117. These markers may identify aberrant PCs in >90% of all patients. For the purpose of MRD detection, cytoplasmic light chain analysis is not required, but combined assessment of cytoplasmic light chain may further improve sensitivity in selected cases when patterns of surface markers alone are equivocal. Clonal shift may also be observed after therapy; consequently, assessment of aberrancies should not be confined to those observed only in the original immunophenotypic profiles. With simultaneous evaluation of multiple parameters, software with the capacity for principal component analysis (PCA) provides improved multidimensional study and more robust classification of different cell types mixed in a sample than conventional visual study of two-dimensional dot plots.

In FCM analysis, the sensitivity of an assay, or limit of detection (LOD), is related to the total number of events acquired. To achieve a sensitivity level of <0.001% or 10^5, at least 3×10^6 events must be acquired. The number of aberrant cells needed to achieve a sensitivity of 10^5 is usually 20 abnormal cells in 2 million events.

As quantitation of aberrant PCs is a critical aspect of MRD assessment, the first step is to draw a broader gate initially and then to exclude nonspecific staining and non-PCs by sequential gating. As described above, gating is based on simultaneous assessment of CD38, CD138, and CD45 expression and light scatter property. The ICCS and ESCCA consensus also recommends that each report specify the level of sensitivity (e.g., LOD and LLOQ) of the assay, the absolute number and percentage of aberrant PCs and their immunophenotypic features, in addition to PC viability and the quality of samples [21].

FCM as a Differential Diagnosis

In addition to distinguishing between benign and malignant PCs, the markers outlined in the ICCS and ESCCA consensus panel described above are also useful for differential diagnosis between PCNs and B cell lymphomas with plasmacytic differentiation, such as lymphoplasmacytic lymphoma associated with Waldenstrom macroglobulinemia (WM). As described earlier, clonal PCs from PCNs differ immunophenotypically from those from WM (Fig. 5.3) and other clonal B cell lymphoproliferative disorders. In lymphoma, the PCs are positive for CD138, but also CD19 and CD45. They may express bright CD38 but are weakly positive for CD138. They may also express surface immunoglobulin light chain [23, 24]. CD56 is usually absent or only rarely expressed in a subset of PCs. Expression of CD117 essentially rules out the possibility of lymphoma except for plasmablastic lymphoma. The PCs in lymphoma also have a low SSC-A compared to the PCs in myeloma. Lymphoma PCs may also express CD20 but are usually negative for CD22. They also express CD27 and CD81.

Immunophenotypic distinction between cases of plasmablastic lymphoma (PL) and myeloma can be difficult sometimes, particularly for cases that do not show aberrant expression of CD56 or CD117. Although cases of plasmablastic lymphoma usually express moderate to bright CD45, this is not invariably the case. *In situ* hybridization for EBV-encoded RNA and immunohistochemical detection of cyclin D1 may help to confirm the diagnosis of PL and myeloma, respectively.

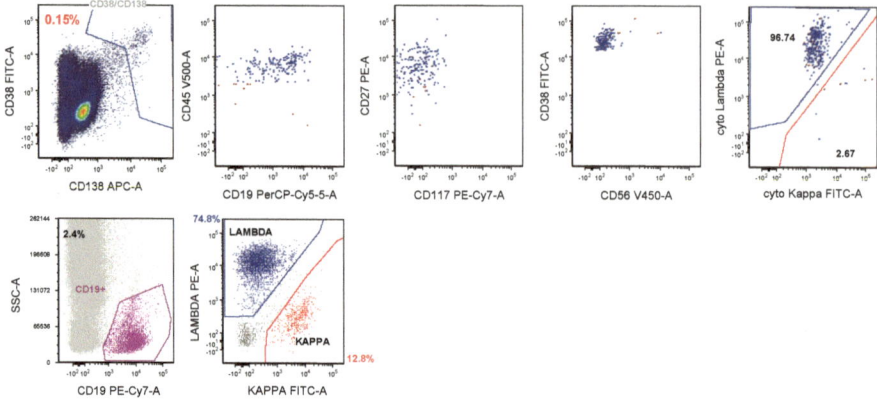

Fig. 5.3 A case illustrates the immunophenotypic differences between plasma cells in lympho-plasmacytic lymphoma versus myeloma: The plasma cells in lymphoma show a low SSC-A and are positive for CD19, CD27, CD45, and cytoplasmic λ light chain restricted. They are negative for CD56 and CD117. CD19+ B cells are also surface λ light chain restricted

FCM for Identification of Therapeutic Target

In recent years, immunotherapy has improved clinical outcomes in myeloma and is recognized as widely applicable regardless of underlying genetic aberrations. Monoclonal antibodies directed against CD38, CD20 and CD138 are being incorporated into the therapeutic armamentarium for myeloma, as part of both frontline and salvage therapies. Antibodies targeting other molecules, such as CD56 and CD117, are currently in development. Given its capacity to both confirm and quantitate the expression of these and other novel markers by neoplastic PCs, FCM will undoubtedly play an important role in guiding the appropriate use of mono-clonal antibody therapy in patients with myeloma.

Summary

In summary, FCM has increasingly become a powerful tool for diagnosis and prognostication of patients with PCNs and related diseases. Data derived from quantification and characterization of aberrant and normal PCs by FCM are useful for differential diagnosis between MGUS and myeloma, assessing the risk of transformation of MGUS or smoldering myeloma to symptomatic myeloma, and for risk stratification among symptomatic patients. Analysis of MRD levels can be used to monitor efficacy of therapy and predict patient outcome. They are therefore effective surrogate that can be used as endpoints for clinical trials, allowing more expedient drug approval by regulatory bodies. With more sensitive methods and analysis software, a future toward a more automated detection and tracking of aberrant cell populations based on comparison of patients' samples against an established reference database of benign and neoplastic PCs is possible. FCM will become an even more powerful tool for diagnosis and management of patients with plasma cell neoplasms in the forseeable future.

References

1. Flores-Montero J, de Tute R, Paiva B, et al. Immunophenotype of normal vs. myeloma plasma cells: toward antibody panel specifications for MRD detection in multiple myeloma. Cytom Part B Clin Cytom. 2015.
2. Liu D, Lin P, Hu Y, et al. Immunophenotypic heterogeneity of normal plasma cells: comparison with minimal residual plasma cell myeloma. J Clin Pathol. 2012;65:823–9.
3. Tembhare PR, Yuan CM, Venzon D, et al. Flow cytometric differentiation of abnormal and normal plasma cells in the bone marrow in patients with multiple myeloma and its precursor diseases. Leuk Res. 2014;38:371–6.
4. Mateo G, Montalban MA, Vidriales MB, et al. Prognostic value of immunophenotyping in multiple myeloma: a study by the PETHEMA/GEM cooperative study groups on patients uniformly treated with high-dose therapy. J Clin Oncol off J Am Soc Clin Oncol. 2008;26:2737–44.
5. Olteanu H, Harrington AM, Kroft SH. CD200 expression in plasma cells of nonmyeloma immunoproliferative disorders: clinicopathologic features and comparison with plasma cell myeloma. Am J Clin Pathol. 2012;138:867–76.
6. Douds JJ, Long DJ, Kim AS, Li S. Diagnostic and prognostic significance of CD200 expression and its stability in plasma cell myeloma. J Clin Pathol. 2014;67:792–6.
7. Alapat D, Coviello-Malle J, Owens R, et al. Diagnostic usefulness and prognostic impact of CD200 expression in lymphoid malignancies and plasma cell myeloma. Am J Clin Pathol. 2012;137:93–100.
8. Lopez-Corral L, Gutierrez NC, Vidriales MB, et al. The progression from MGUS to smoldering myeloma and eventually to multiple myeloma involves a clonal expansion of genetically abnormal plasma cells. Clin Cancer Res Off J Am Assoc Cancer Res. 2011;17:1692–700.
9. Paiva B, Perez-Andres M, Vidriales MB, et al. Competition between clonal plasma cells and normal cells for potentially overlapping bone marrow niches is associated with a progressively altered cellular distribution in MGUS vs myeloma. Leukemia. 2011;25:697–706.
10. Paiva B, Vidriales MB, Mateo G, et al. The persistence of immunophenotypically normal residual bone marrow plasma cells at diagnosis identifies a good prognostic subgroup of symptomatic multiple myeloma patients. Blood. 2009;114:4369–72.
11. Paiva B, Vidriales MB, Rosinol L, et al. A multiparameter flow cytometry immunophenotypic algorithm for the identification of newly diagnosed symptomatic myeloma with an MGUS-like signature and long-term disease control. Leukemia. 2013;27:2056–61.
12. Paiva B, Gutierrez NC, Chen X, et al. Clinical significance of CD81 expression by clonal plasma cells in high-risk smoldering and symptomatic multiple myeloma patients. Leukemia. 2012;26:1862–9.
13. Mateo G, Castellanos M, Rasillo A, et al. Genetic abnormalities and patterns of antigenic expression in multiple myeloma. Clin Cancer Res Off J Am Assoc Cancer Res. 2005;11:3661–7.
14. Bataille R, Jego G, Robillard N, et al. The phenotype of normal, reactive and malignant plasma cells. Identification of "many and multiple myelomas" and of new targets for myeloma therapy. Haematologica. 2006;91:1234–40.

15. Paino T, Paiva B, Sayagues JM, et al. Phenotypic identification of subclones in multiple myeloma with different chemoresistant, cytogenetic and clonogenic potential. Leukemia. 2015;29:1186–94.
16. Gonsalves WI, Rajkumar SV, Gupta V, et al. Quantification of clonal circulating plasma cells in newly diagnosed multiple myeloma: implications for redefining high-risk myeloma. Leukemia. 2014;28:2060–5.
17. Rajkumar SV, Harousseau JL, Durie B, et al. Consensus recommendations for the uniform reporting of clinical trials: report of the International Myeloma Workshop Consensus Panel 1. Blood. 2011;117:4691–5.
18. Rawstron AC, Child JA, de Tute RM, et al. Minimal residual disease assessed by multiparameter flow cytometry in multiple myeloma: impact on outcome in the Medical Research Council Myeloma IX Study. J Clin Oncol Off J Am Soc Clin Oncol. 2013;31:2540–7.
19. Rawstron AC, Gregory WM, de Tute RM, et al. Minimal residual disease in myeloma by flow cytometry: independent prediction of survival benefit per log reduction. Blood. 2015;125:1932–5.
20. Paiva B, Gutierrez NC, Rosinol L, et al. High-risk cytogenetics and persistent minimal residual disease by multiparameter flow cytometry predict unsustained complete response after autologous stem cell transplantation in multiple myeloma. Blood. 2012;119:687–91.
21. Arroz M, Came N, Lin P, et al. Consensus guidelines on plasma cell myeloma minimal residual disease analysis and reporting. Clinical cytometry: Cytom Part B; 2015.
22. Stetler-Stevenson M, Paiva B, Stoolman L, et al. Consensus guidelines for myeloma minimal residual disease sample staining and data acquisition. Cytome Part B Clin Cytom. 2015.
23. Morice WG, Chen D, Kurtin PJ, Hanson CA, McPhail ED. Novel immunophenotypic features of marrow lymphoplasmacytic lymphoma and correlation with Waldenstrom's macroglobulinemia. Modern pathology: an official journal of the United States and Canadian Academy of Pathology, Inc 2009;22:807–16.
24. Howard MT, Hodnefield J, Morice WG. Immunohistochemical phenotyping of plasma cells in lymphoplasmacytic lymphoma/Waldenstrom's macroglobulinemia is comparable to flow cytometric techniques. Clin Lymphoma Myeloma Leuk. 2011;11:96–8.

Chapter 6
Renal Manifestations of Plasma Cell Neoplasms

L. Nicholas Cossey and Shree G. Sharma

Introduction

Dysproteinemia is an important disease entity characterized by excessive amounts of abnormal monoclonal proteins in the circulation. The patient may be asymptomatic or may have symptoms secondary to deposition in the organs. The symptoms depend on the organ involved and the severity of involvement. If the patient stays asymptomatic for a long period, disease may become widespread and the deposition of the abnormal protein may be extensive. It is not only important to diagnose the type of dysproteinemia, but is also important to find out the cause of dysproteinemia for appropriate treatment. Common clinical causes of dysproteinemia are plasma cell myeloma, monoclonal gammopathy of unknown significance (MGUS), primary amyloidosis, and lymphoproliferative disorders.

The most common dysproteinemia-related kidney diseases include light chain cast nephropathy, light chain proximal tubulopathy (LCPT), amyloidosis, monoclonal immunoglobulin deposit diseases (MIDD) and a recently described entity known as monoclonal gammopathy of renal significance (MGRS), in which the patient's dysproteinemia may present only with renal dysfunction. Plasma cell myeloma is one of the most common causes of dysproteinemia and frequently affects the kidney. Greater than 50% of the patients with plasma cell myeloma present with renal insufficiency at the time of diagnosis [1]. Renal biopsy is an important tool in these cases to confirm involvement of kidney. Renal biopsy interpretation can be challenging in subtle cases and at initial stages of the disease. The present chapter will discuss important findings seen on renal biopsy and their interpretation.

L. Nicholas Cossey · S.G. Sharma (✉)
Nephropath, 10810 Executive Center Dr. Ste. 100, Little Rock, AR 72211, USA
e-mail: nich.cossey@nephropath.com

© Springer International Publishing Switzerland 2016
R.B. Lorsbach and M. Yared (eds.), *Plasma Cell Neoplasms*,
DOI 10.1007/978-3-319-42370-8_6

Light Chain Cast Nephropathy

Light chain cast nephropathy (a.k.a. myeloma cast nephropathy) commonly presents as acute kidney injury and is one of the most common manifestations of plasma cell myeloma. The patient often has non-nephrotic range proteinuria that is not detected by urine dipstick. This is due to the fact that dipstick detects albumin but the proteinuria in these patients is primarily due to light chains (a.k.a. Bence-Jones protein) [2]. On histopathology, the disease is characterized by formation of casts in the distal tubules of the kidney. The casts are formed by the combination of Tamm–Horsfall protein and monoclonal light chains and are brittle, which imparts a fractured cast appearance [3]. The casts can lead to both obstruction as well as cause direct toxicity to the tubules. In some cases, the patient has an identifiable precipitating factor such as dehydration, diuretics, hypercalcemia, infections, and nephrotoxins [2]. Renal biopsy is necessary to confirm the diagnosis.

Light microscopy: The pathology is most prominently seen in the tubulointerstitial compartment. Abnormal casts are identified predominantly in the distal nephron segments. The casts appear eosinophilic with H&E stain (Fig. 6.1), pale on PAS stain (Fig. 6.2) and are polychromatic (red and blue) on the trichrome stain (Figs. 6.3 and 6.4). The casts have variable geometric shapes and may have a sharp-edged, angulated or fractured appearance (Figs. 6.5, 6.6, 6.7, 6.8, 6.9, 6.10 and 6.11). The casts may also be surrounded by epithelial cells, giant cells or inflammatory cells (Figs. 6.12 and 6.13) and can have a crystalline appearance (Figs. 6.14 and 6.15). These findings may be accompanied by interstitial inflammation. This diagnosis is easier when the casts are numerous and may be challenging in cases with only occasional casts. It is important to pay careful attention to the morphology and staining characteristics in these cases (Fig. 6.16). In cases with only a few casts visible by light microscopy, performing paraffin-retrieved immunofluorescence (IF) for kappa and lambda light chains with an additional PAS stain on the slide in between those slides (Figs. 6.17, 6.18 and 6.19) can be very helpful in proving the monoclonality of the casts. Rarely casts can be positive for Congo red stain and show apple-green birefringence under polarized light (Figs. 6.20, 6.21 and 6.22). These casts may also exhibit a fibrillar appearance on electron microscopy [4]. Importantly, these cases are not associated with systemic amyloidosis and this finding should not be interpreted to represent renal amyloidosis.

Immunofluorescence: The casts are restricted for either kappa or lambda light chain (Figs. 6.23 and 6.24). Kappa is more commonly seen than lambda. In cases with few casts, performing IF on paraffin tissue can be very helpful (Figs. 6.25 and 6.26).

Electron microscopy: Electron microscopy reveals a variable appearance of the casts with nonspecific substructure formation. The monoclonality of the casts can be demonstrated by immunogold labeling.

Fig. 6.1 Atypical eosinophilic crystalline casts (H&E stain, original magnification, 400×)

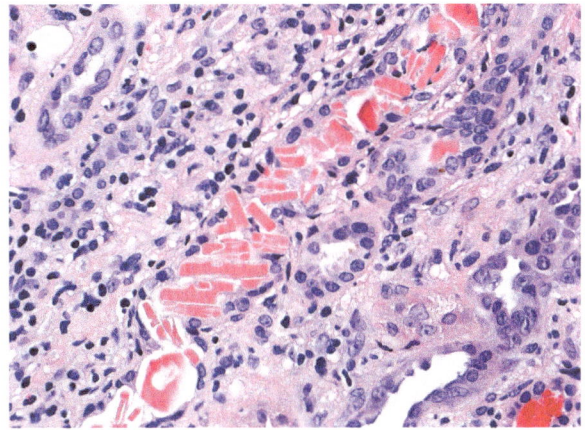

Fig. 6.2 PAS-pale casts (PAS stain, original magnification, 400×)

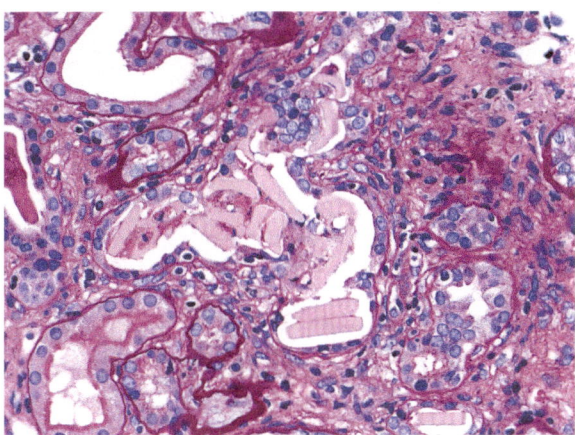

Fig. 6.3 Polychromatic cast (Masson trichrome stain, original magnification, 400×)

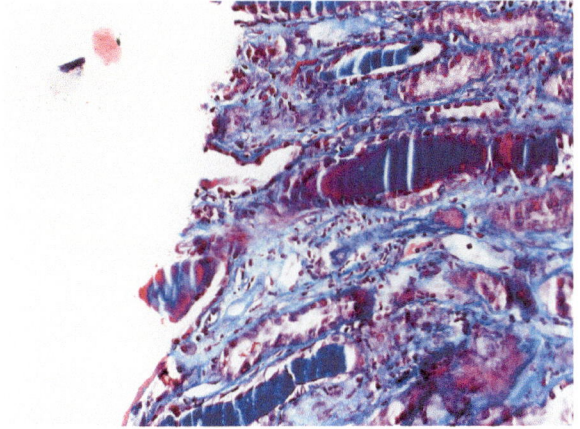

Fig. 6.4 Polychromatic cast
(Massson trichrome stain,
original magnification, 400×)

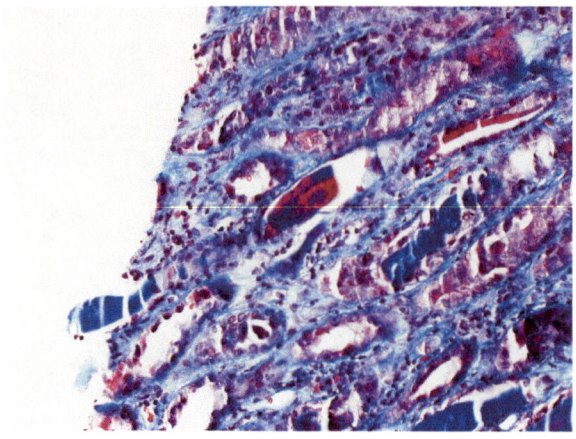

Fig. 6.5 Variable appearance
of the atypical casts (H&E
stain, original magnification,
400×)

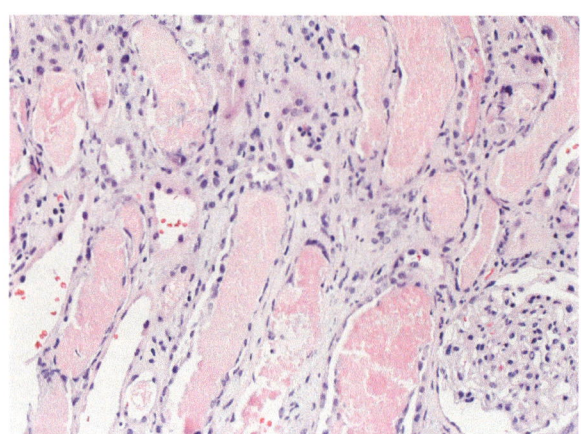

Fig. 6.6 Variable appearance
of the atypical casts (H&E
stain, original magnification,
400×)

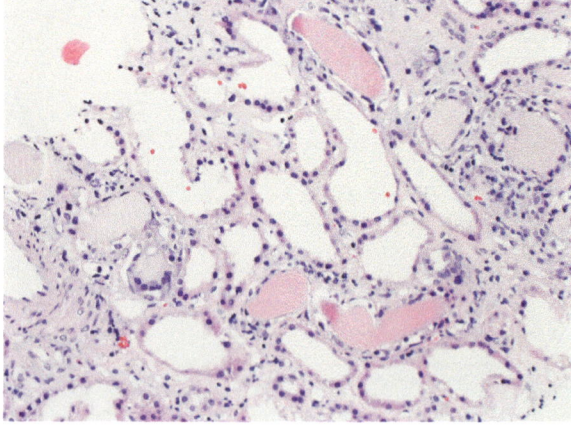

Fig. 6.7 Variable appearance of the atypical casts (PAS stain, original magnification, 400×)

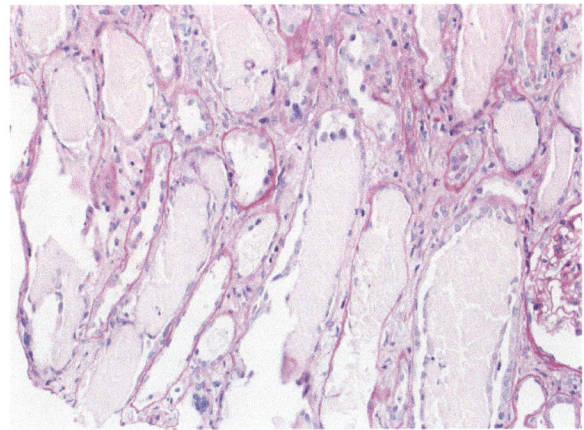

Fig. 6.8 Variable appearance of the atypical casts (PAS stain, original magnification, 400×)

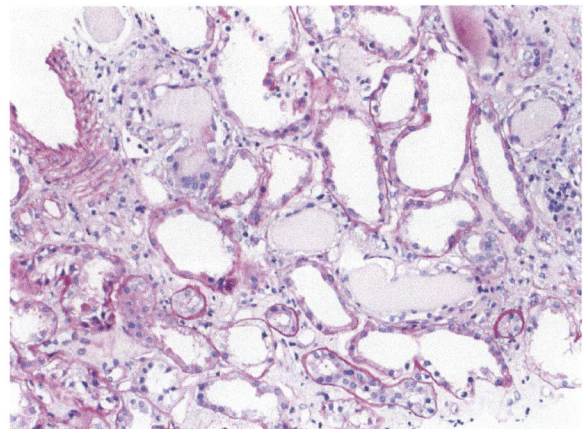

Fig. 6.9 Variable appearance of the atypical casts (Masson's trichrome stain, original magnification, 400×)

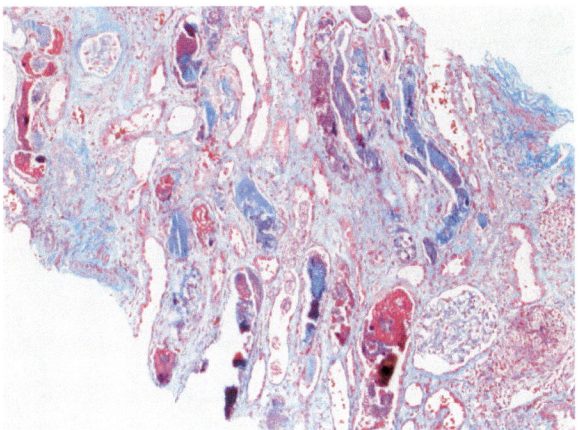

Fig. 6.10 Variable
appearance of the atypical
casts (Masson's trichrome
stain, original magnification,
400×)

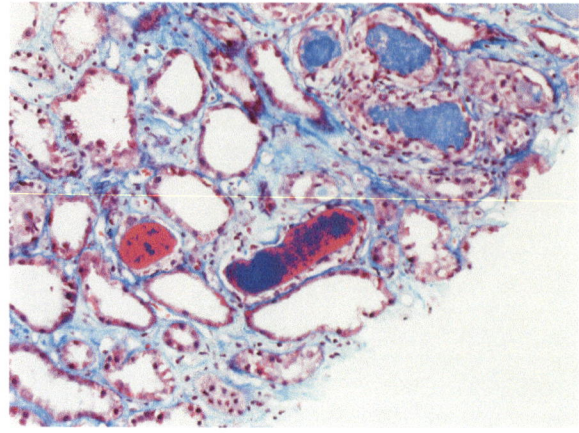

Fig. 6.11 Variable
appearance of the atypical
casts (Jones Methenamine
Silver stain, original
magnification, 400×)

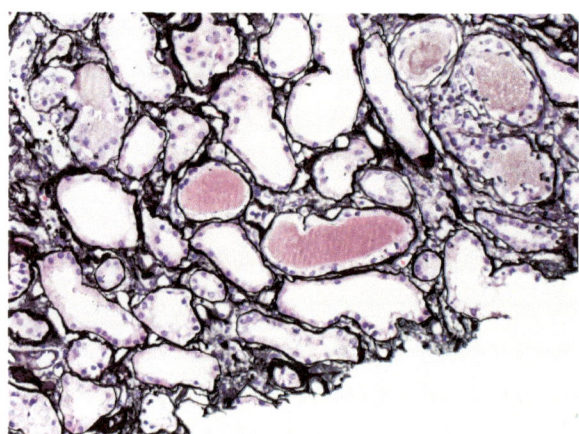

Fig. 6.12 Eosinophilic casts
with surrounding reaction
(H&E stain, original
magnification, 400×)

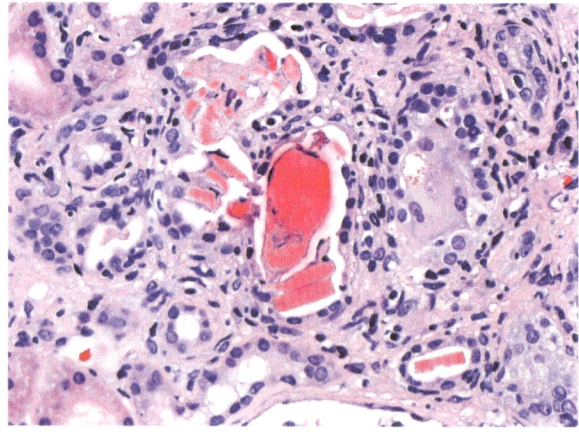

Fig. 6.13 PAS-pale casts with surrounding reaction (PAS stain, original magnification, 400×)

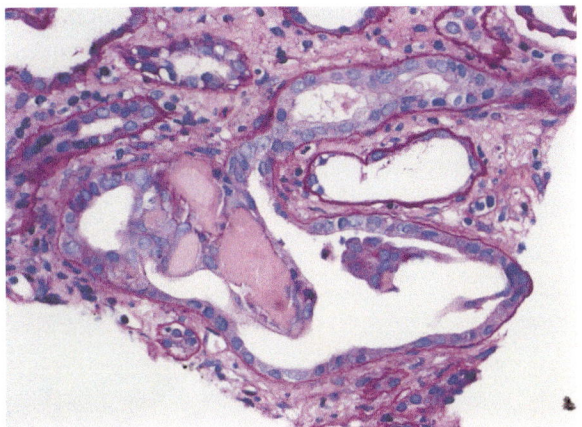

Fig. 6.14 Crystalline casts (Jones Methenamine Silver stain, original magnification, 400×)

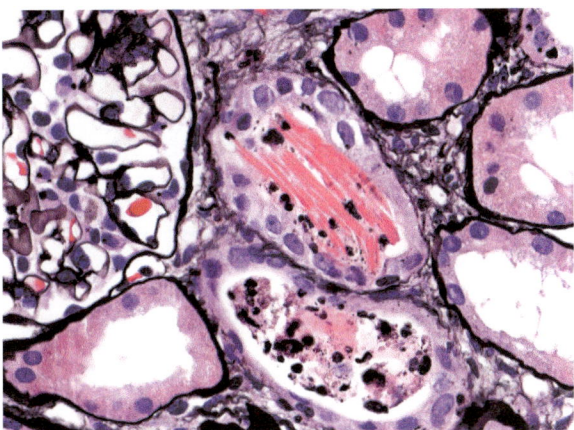

Fig. 6.15 Crystalline casts (Masson's trichrome stain, original magnification, 400×)

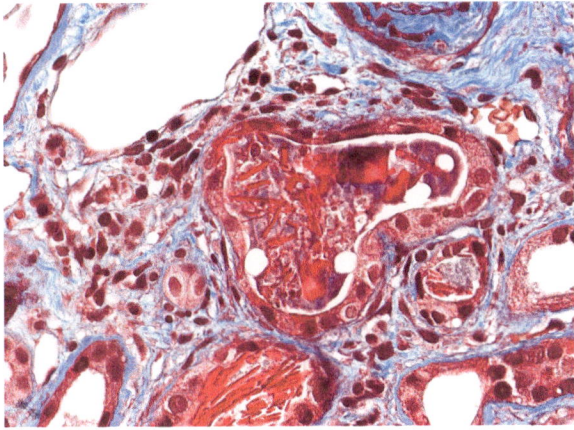

Fig. 6.16 Atypical PAS-pale myeloma casts seen in the background and PAS-positive hyaline cast in the center (PAS stain, original magnification 400×)

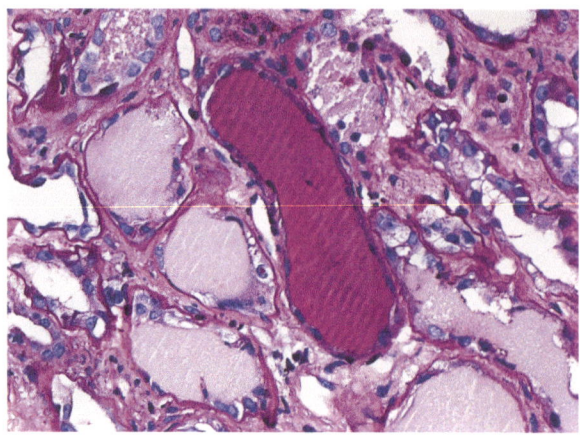

Fig. 6.17 Casts positive for kappa light stain (Parrafin-IF, original magnification, 400×)

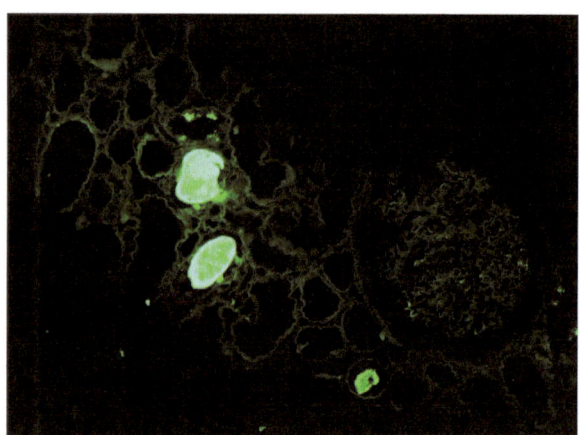

Fig. 6.18 PAS-pale casts (PAS stain, original magnification, 400×)

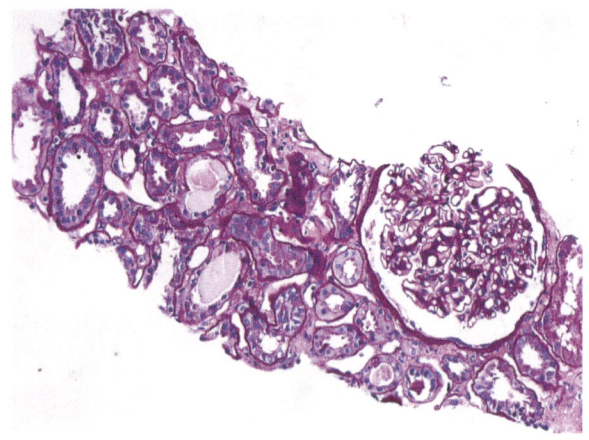

Fig. 6.19 Casts negative for lambda light stain (Parrafin-IF, original magnification, 400×)

Fig. 6.20 Myeloma cast surrounded by inflammatory and giant cells (PAS stain, original magnification, 400×)

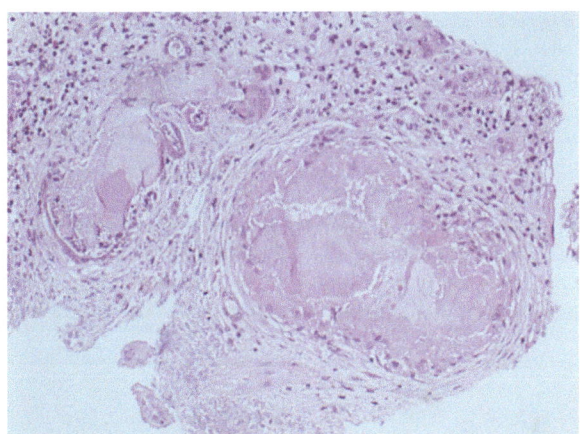

Fig. 6.21 Myeloma cast showing peripheral Congo *red* positivity (Congo *red* stain, original magnification, 400×)

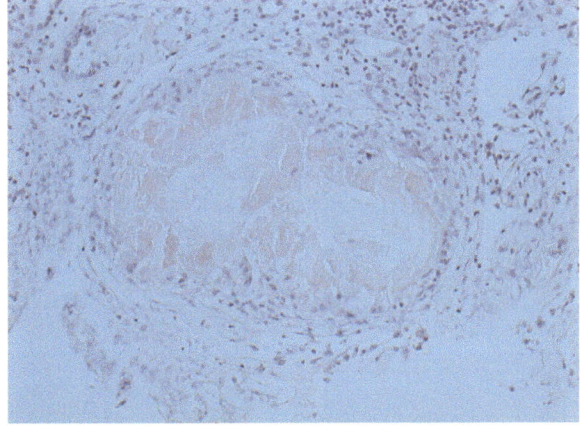

Fig. 6.22 Myeloma cast
showing apple-green
birefringence under polarized
light (Congo *red* stain,
original magnification, 400×)

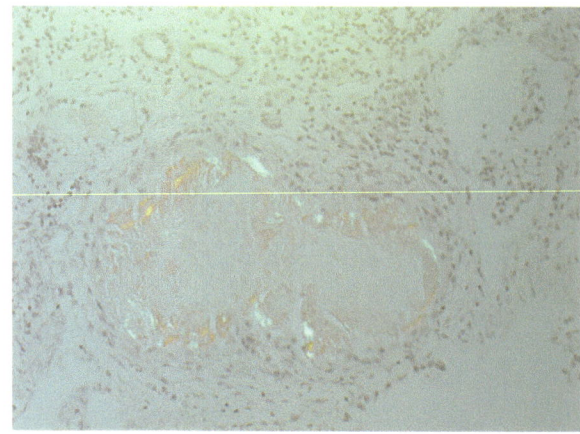

Fig. 6.23 Myeloma casts
showing diffuse 3+ positivity
for kappa light chain (IF,
original magnification, 400×)

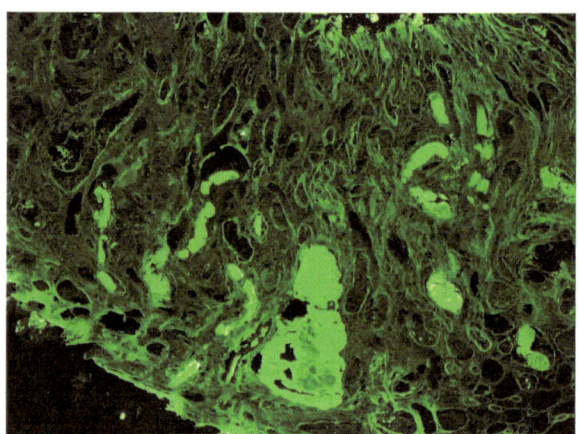

Fig. 6.24 Myeloma casts are
negative for lambda light
chain (IF, original
magnification, 400×)

Fig. 6.25 IF performed on paraffin tissue showing kappa light chain positive casts (original magnification, 400×)

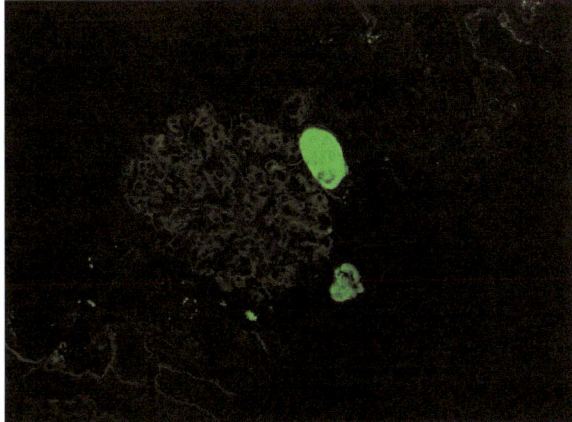

Fig. 6.26 IF performed on paraffin tissue showing lambda negative casts (original magnification, 400×)

Light Chain Proximal Tubulopathy

Light chain proximal tubulopathy (LCPT) can occur in isolation or in combination with other dysproteinemia-related renal diseases. When present in isolation, the presentation can vary from acute kidney injury to slow rise in creatinine. When LCPT presents as Fanconi syndrome (normoglycemic glycosuria, aminoaciduria, uricosuria, hyperphosphaturia with hypophosphatemia and type II renal tubular acidosis), it is known as light chain Fanconi syndrome. In LCPT, the light chains absorbed by the proximal tubules are not digested by the lysosomes, resulting in their accumulation and crystallization within the tubules and subsequent damage [5]. At the time of renal biopsy, there is often no history of dysproteinemia or plasma cell myeloma.

Light microscopy: On light microscopic examination, hematoxylin and eosin stain shows features of tubular injury and needle-shaped empty spaces in cases with crystal formation. In cases without crystal formation, cytoplasm is pink and granular due to the presence of lysosomes [6–8]. The crystals and inclusions are PAS-pale and stain strongly pink with the trichrome stain (Figs. 6.27, 6.28, 6.29 and 6.30). The crystals are more easily visualized on the survey sections for electron microscopy by toluidine blue stain (Figs. 6.31 and 6.32). Glomeruli are generally unremarkable in the absence of superimposed glomerular disease. Rarely, eosinophilic cytoplasmic droplets can be seen scattered in the podocytes, parietal cells, distal, and collecting tubules and in the interstitium [9].

Immunofluorescence: IF shows monotypic staining for either kappa or lambda light chain (Figs. 6.33 and 6.34). Cases with crystal formation are typically kappa-restricted while lambda-restricted cases are less prone to form crystals.

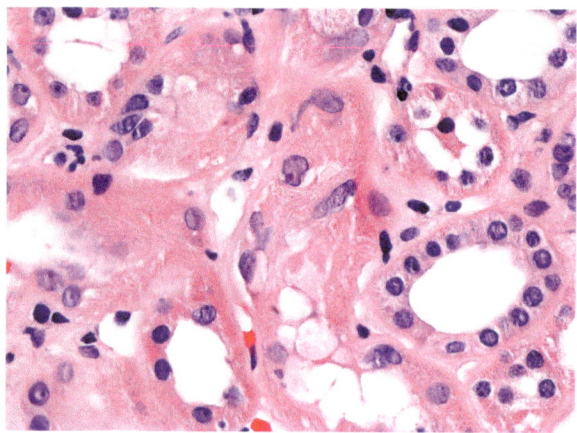

Fig. 6.27 Needle shaped crystals on H&E stain (original magnification, 400×)

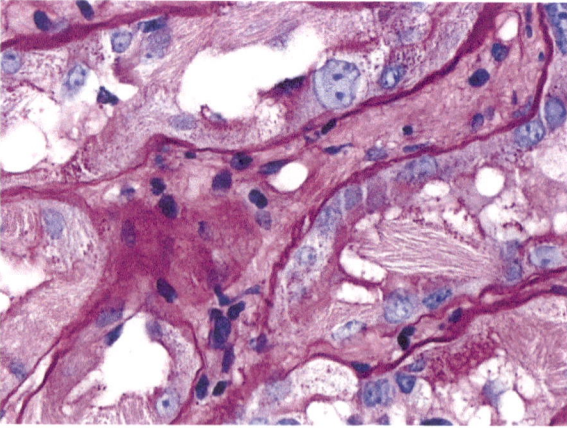

Fig. 6.28 Needle shaped PAS-pale crystals (original magnification, 400×)

Fig. 6.29 Fuchsinophilic cytoplasm on Masson's trichrome stain (original magnification, 400×)

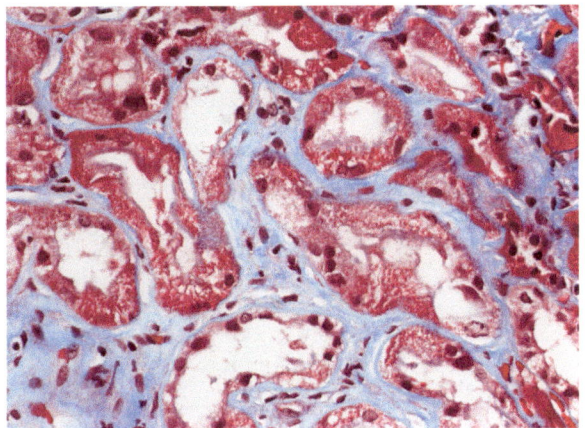

Fig. 6.30 PAS pale cytoplasm (original magnification, 400×)

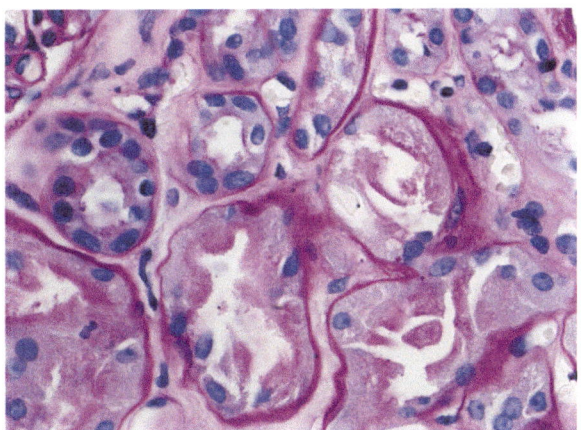

Fig. 6.31 Crystals seen on toluidine blue stain sections prepared for electron microscopy (original magnification, 200×)

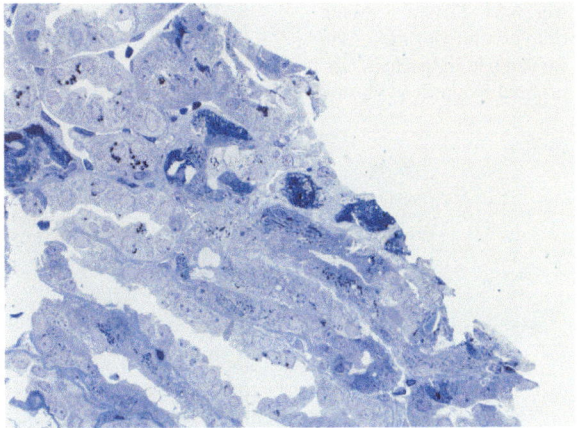

Fig. 6.32 Crystals seen on toluidine blue stain sections prepared for electron microscopy (original magnification, 400×)

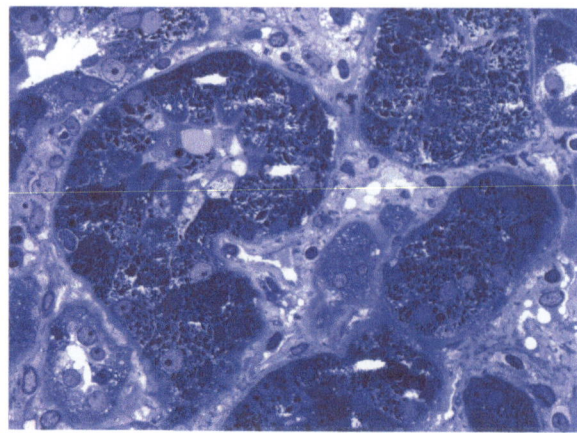

Fig. 6.33 Proximal tubule droplets showing positivity for kappa light chain (IF, original magnification, 400×)

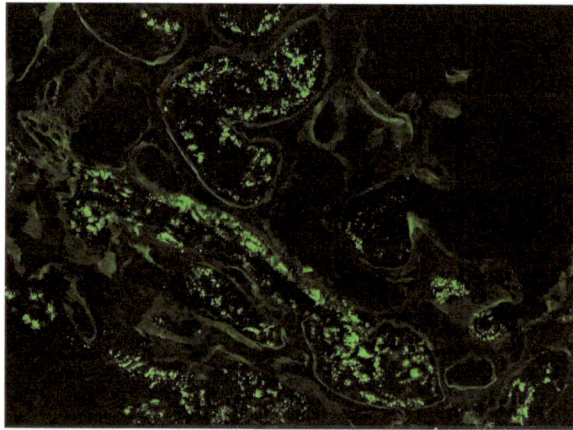

Fig. 6.34 Proximal tubule droplets showing negativity for lambda light chain (IF, original magnification, 400×)

Fig. 6.35 Proximal tubule showing kappa light chain positive crystals on IF performed on paraffin tissue (original magnification, 400×)

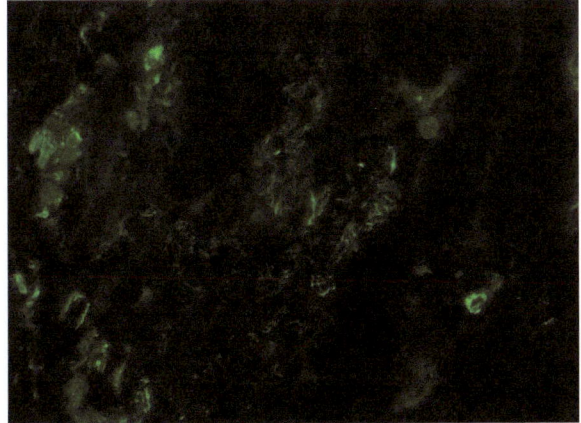

Fig. 6.36 Proximal tubule showing lambda light chain negative crystals on IF performed on paraffin tissue (original magnification, 400×)

IF performed on paraffin-retrieved tissue increases the sensitivity of diagnosis (especially in crystal forming cases) and should always be performed if the suspicion is high and the routine IF is negative (Figs. 6.35 and 6.36) [10].

Electron microscopy: Proximal tubules display needle or rhomboid shaped crystals in cases with crystal formation (Figs. 6.37, 6.38, 6.39 and 6.40) while distended lysosomes can be seen in cases without crystal formation (Figs. 6.41 and 6.42). The lysosomes can be highlighted by immunolabeling to reveal a monoclonal light chain. Immunolabeling is not widely available; therefore, in practice the diagnosis is made on the basis of immunofluorescence, light microscopy, and electron microscopic features.

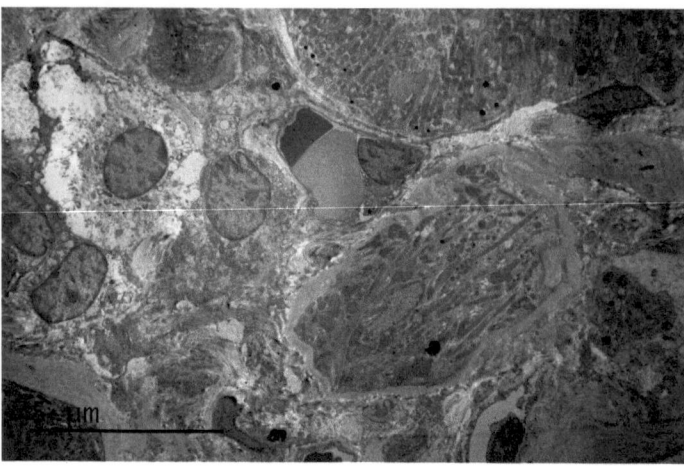

Fig. 6.37 Needle shaped crystals seen on electron microscopy (original magnification, 10,000×)

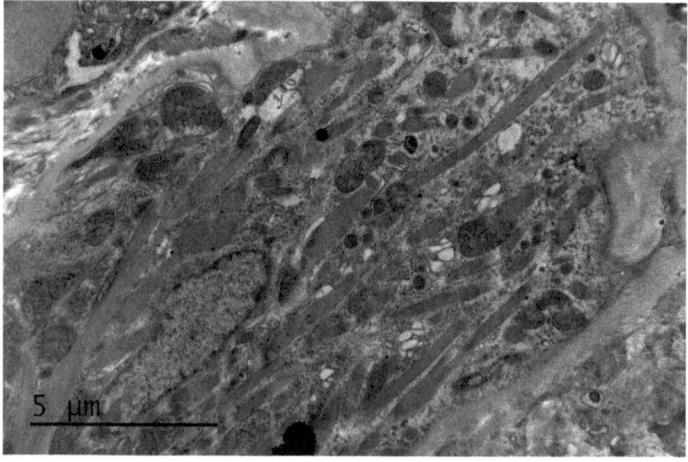

Fig. 6.38 Needle shaped crystals seen on electron microscopy (original magnification, 25,000×)

Monoclonal Ig Deposition Disease

Monoclonal Ig deposition disease (MIDD) is characterized by deposition of monoclonal proteins (light and/or heavy chains). Deposition of the abnormal protein may occur in multiple organs including kidney, heart, liver, gastrointestinal tract, blood vessels, skin, lung, salivary glands and nerves. The clinical presentation depends on the predominant organ involved. In the kidney, depending on the extent and the pattern of involvement (predominantly tubulointerstitial versus glomerular),

Fig. 6.39 Rhomboidal crystals seen on electron microscopy (original magnification, 15,000×)

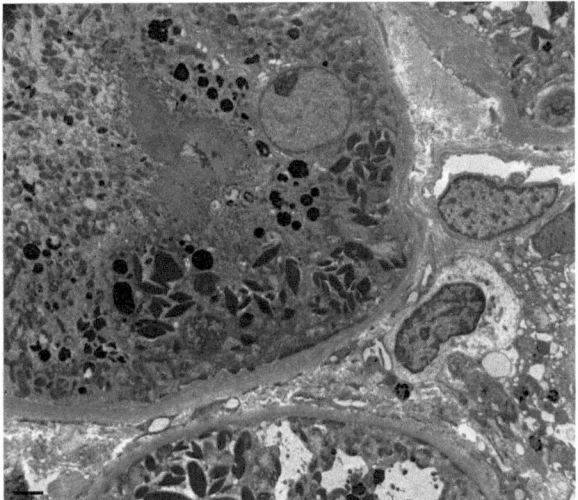

Fig. 6.40 Rhomboidal crystals seen on electron microscopy (original magnification, 25,000×)

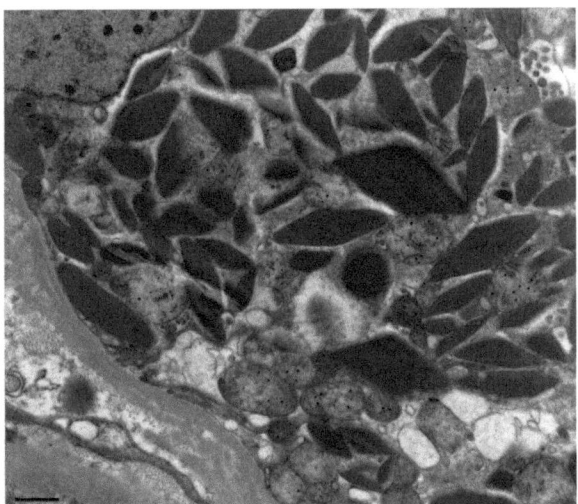

the patient may present with subnephrotic to nephrotic range proteinuria. Laboratory evaluation reveals evidence of monoclonality in approximately 80% of patients with light chain deposition disease (LCDD) and 60% of patients with heavy chain deposition disease (HCDD). Besides features of dysproteinemia, patients with HCDD and light- and heavy-chain deposition disease (LHCDD) may also present with hypocomplementemia because of the complement fixing ability of the heavy chains [11–13].

Light microscopy: Glomerular lesions are very heterogeneous, displaying expansion of the mesangium, often giving a characteristic nodular appearance

Fig. 6.41 Tubular cytoplasm
showing phagolysosomes
with no crystals (original
magnification, 15,000×)

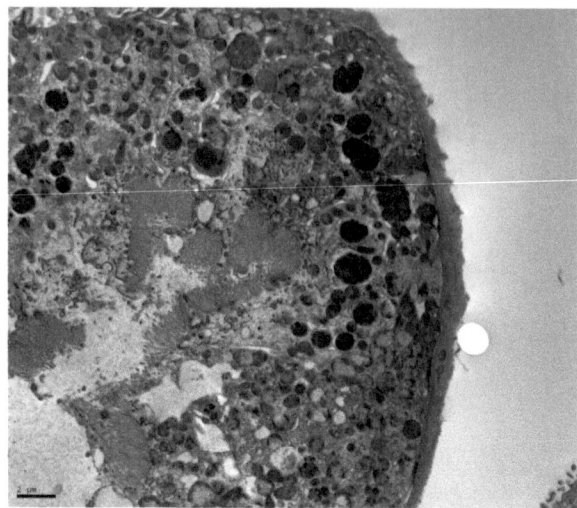

Fig. 6.42 Tubular cytoplasm
showing phagolysosomes
with no crystals (original
magnification, 30,000×)

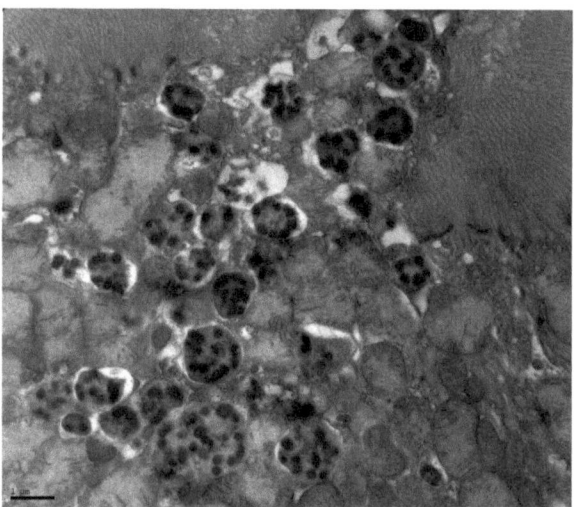

(Fig. 6.43). Nodular mesangial expansion is seen more often with HCDD compared
to LCDD and can be confused with diabetic glomerulosclerosis or other causes of
nodular mesangial sclerosis. The glomerular basement membrane is thickened and
can show membranoproliferative features (mesangial interposition and duplication
of the GBM). Tubular basement membranes and vascular basement membranes
also show thickening (Figs. 6.44, 6.45 and 6.46). On light microscopy, the deposits
are eosinophilic by H&E stain, PAS positive, red on trichrome stain, and are weakly
staining or not stained on the silver stain (Fig. 6.47). The staining pattern of the
nodules helps to differentiate MIDD from diabetes on light microscopy, in which

Fig. 6.43 Nodular sclerosing monoclonal immunoglobulin deposition disease (PAS stain, original magnification, 400×)

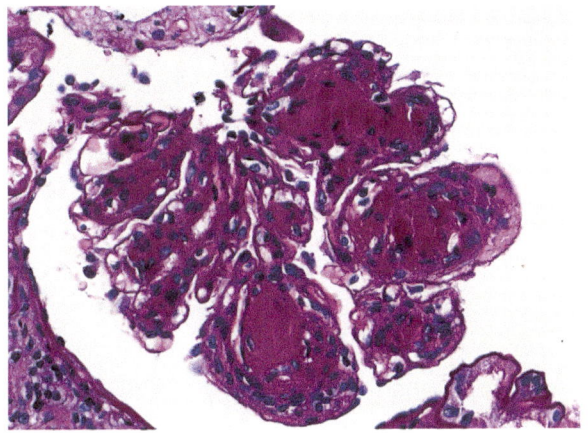

Fig. 6.44 PAS-positive deposits within the blood vessel wall (original magnification, 400×)

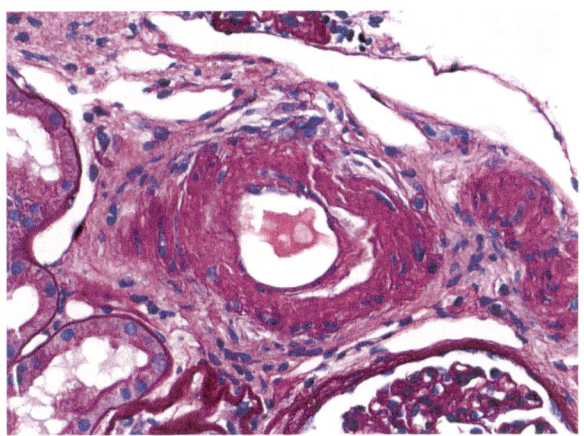

Fig. 6.45 Thickening of the blood vessel wall and TBMs by PAS-positive material (original magnification, 400×)

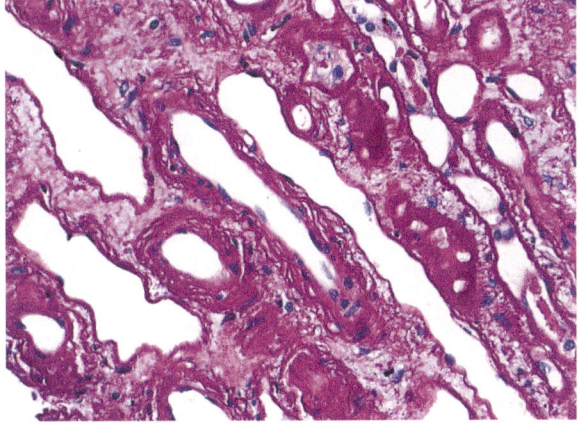

Fig. 6.46 Thickening of the tubular basement membranes by PAS-positive material (original magnification, 400×)

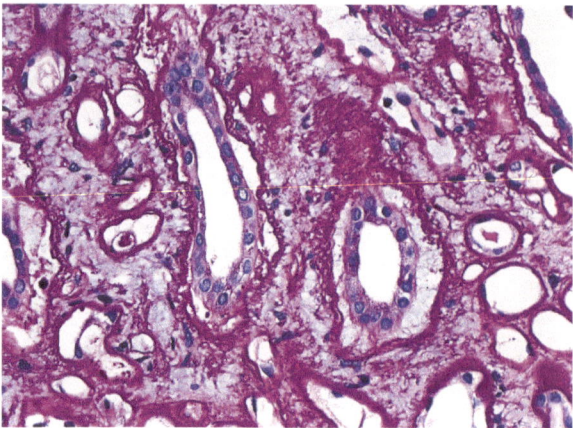

Fig. 6.47 Deposits within the blood vessel wall are non-argyrophilic (original magnification, 400×)

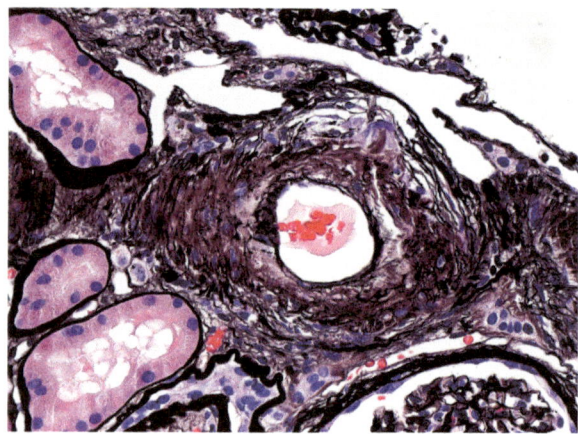

the nodules are strongly silver positive and trichrome blue. Crescents are not commonly seen in cases of LCDD but can be seen in one third of the cases of HCDD. The deposits can sometimes extend to the interstitium and may be associated with giant cell reaction. The deposits in MIDD are Congo red negative.

Immunofluorescence: Immunofluorescence reveals distinct linear staining of the renal basement membranes (tubular basement membrane, glomerular basement membrane and vessel wall) by the involved monotypic light chain (Figs. 6.48, 6.49, 6.50 and 6.51). In LCDD, kappa is more commonly seen than lambda while in HCDD γ is more commonly seen than α and μ. In γ HCDD staining for IgG subtypes (IgG1, IgG2, IgG3, and IgG4) is helpful in confirming the diagnosis. Sometimes, the only finding in cases of LCDD is the linear staining on IF, without corresponding deposits on the electron microscopy. These cases may represent early LCDD or may be secondary to nonspecific binding of the monoclonal protein to the basement membranes [12]. In cases where deposits are present by IF only, it

Fig. 6.48 IF for kappa light chain showing staining of tubular and glomerular basement membranes (original magnification, 200×)

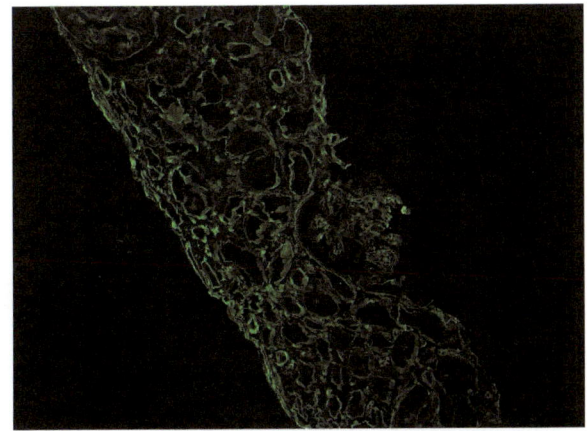

Fig. 6.49 IF showing negative staining for lambda light chain (original magnification, 200×)

Fig. 6.50 IF showing blood vessel wall staining by kappa light chain (original magnification, 400×)

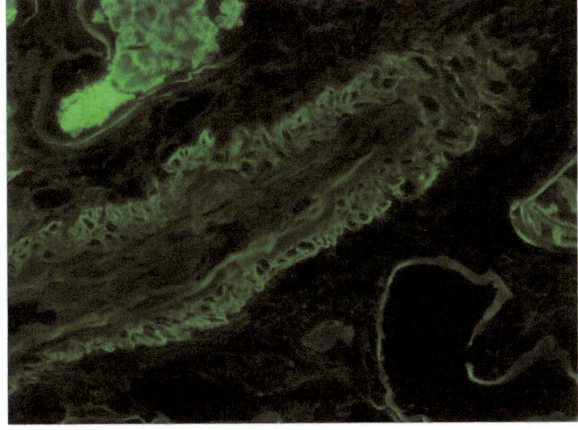

Fig. 6.51 IF showing
negative blood vessel wall
staining for lambda light
chain (original magnification,
400×)

is important to make note of this in the pathology report, so that the patient can be assessed and the report can be interpreted in the appropriate clinical context.

Electron microscopy: Ultrastructurally, renal basement membrane deposits are seen and appear as finely granular, punctate, or powdery (Figs. 6.52, 6.53 and 6.54). The deposits lack substructure formation and are electron-dense. The deposits are more commonly seen toward the inner aspect (subendothelial) of the GBM, outer aspect of the TBM, and within the mesangial matrix, Bowman's capsule and perimyocyte matrix of the blood vessel wall [11].

Fig. 6.52 Fine granular
electron-dense deposits along
the tubular basement
membranes (original
magnification, 4000×)

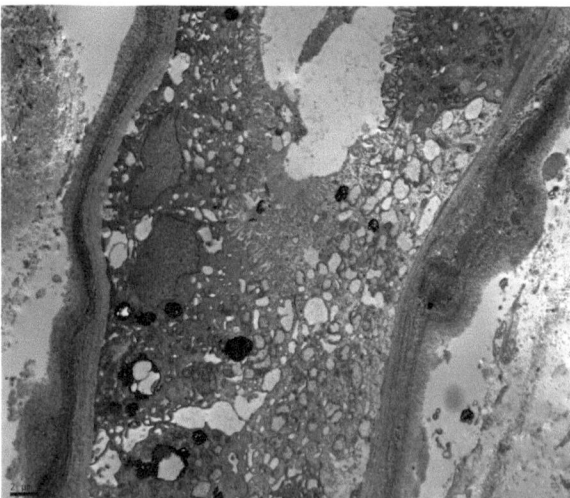

Fig. 6.53 Fine granular electron-dense deposits along the tubular basement membrane (original magnification, 25,000×)

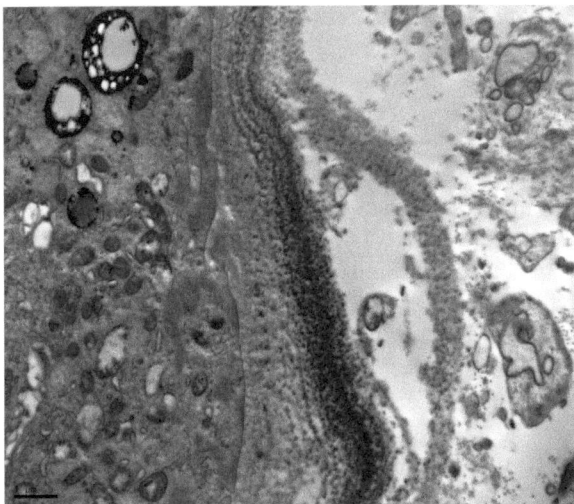

Fig. 6.54 Powdery deposits along GBMs (original magnification, 10,000×)

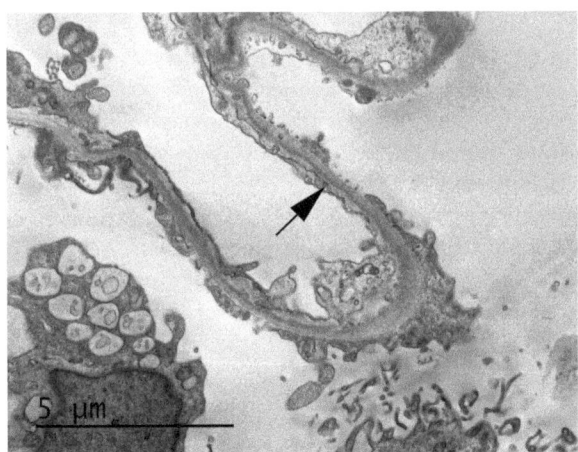

Renal Amyloidosis

Amyloid refers to a group of more than 25 proteins that display abnormal folding due to repeating beta pleated sheet formation [14]. These abnormal proteins are prone to extracellular tissue deposition where they can impair normal physiologic function. Amyloidosis can produce systemic or localized disease and often has specific clinical and/or histologic findings associated with specific subtypes. The kidney is relevant in systemic amyloidosis as it is the organ most commonly affected [14]. Amyloidosis is grouped into subtypes based on the underlying abnormal protein. An accurate subtype designation is paramount for further clinical workup and treatment. The two most commonly encountered types of amyloidosis are the AL type arising from immunoglobulin light chain secretion in plasma cell dyscrasias, and the AA type derived from serum amyloid A most typically associated with underlying chronic inflammation and infection [14]. Other, less common types of amyloid that can affect the kidney include LECT2 (leukocyte chemotactic factor 2) [15, 16], AH (Ig heavy chain), AHL (Ig heavy and light chain), AApo (Apo A I), AII (Apo A II), AIV (Apo A IV) [17], AFib (Fibrinogen A α chain) [18], ATTR (Transthyretin), ALys (Lysozyme), β2-microglobulin and AGel (Gelsolin) [19]. The common types (AL, AA, LECT2) of amyloidosis can be differentiated by morphology, immunofluorescence, and immunohistochemistry, while the less common forms require more sophisticated testing [14].

The clinical manifestations of renal amyloidosis vary depending on subtype and location of deposition; however, the most common presenting symptoms in renal amyloidosis are proteinuria and a variable degree of renal dysfunction. The proteinuria can range from marginal to nephrotic range (>3.5 gm/24 h) and can present as an acute or chronic phenomenon. In the AL subtype, these findings may also be associated with clinical signs and symptoms of plasma cell myeloma such as bone pain, hypercalcemia and anemia [20].

Light Microscopy: All amyloid proteins share a common, basic morphology showing pale, eosinophilic, homogenous material that frequently involves the glomeruli, arteries, arterioles and less commonly the tubulointerstitium (Fig. 6.55). It is identified by positive staining on Congo red special stain with apple-green (may range from yellow/orange/green) birefringence upon polarization (Figs. 6.56, 6.57, 6.58 and 6.59) [14]. Amyloid deposition can range from extensive to very focal and can be easily overlooked in more subtle cases. Amyloid is typically PAS negative, pale blue on Masson's trichrome and has variable argyrophilia on silver stain (Figs. 6.60, 6.61 and 6.62). Peripheral capillary loops will rarely show spicular alterations (Figs. 6.63 and 6.64). While some subtypes of amyloid have well described morphologic patterns, most have a nonspecific pattern and require

Fig. 6.55 Pale, eosinophilic amyloid material within the mesangium and capillary walls (H&E stain, original magnification, 200×)

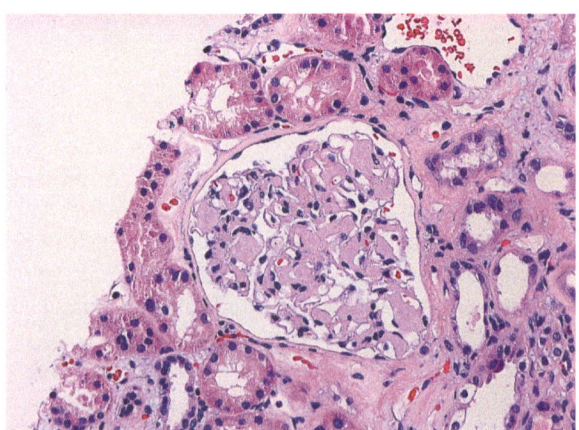

Fig. 6.56 Congo *red* stain showing positive staining with glomeruli (original magnification, 400×)

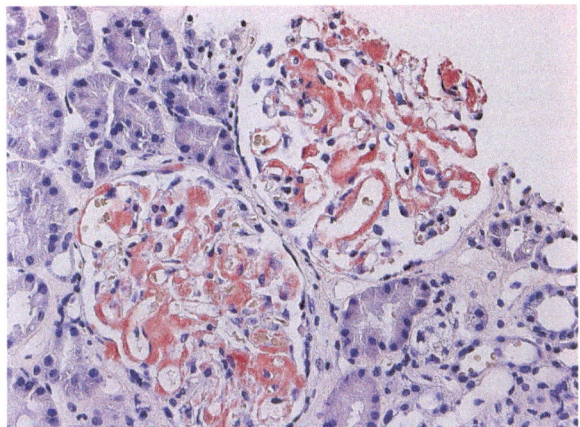

Fig. 6.57 Congo *red* stain showing positive staining within an artery (original magnification, 100×)

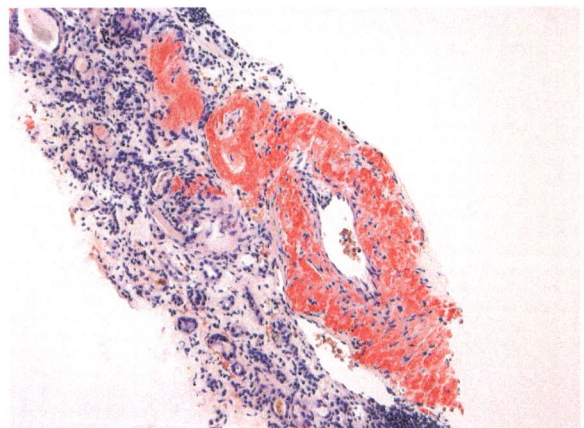

Fig. 6.58 Apple *green* to *yellow* birefringence on polarization within glomeruli (Congo *red* stain, original magnification 400×)

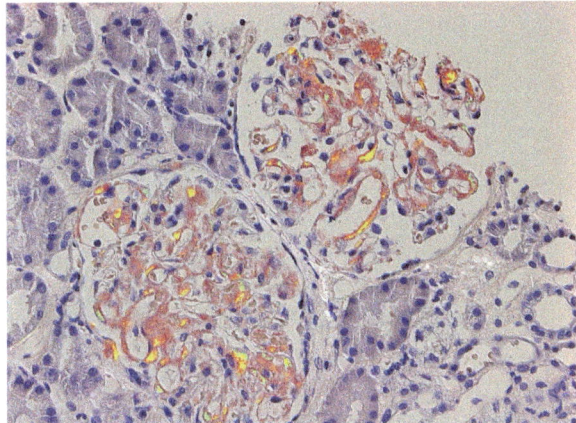

Fig. 6.59 Apple *green* to *yellow/orange* birefringence on polarization within an artery (Congo *red* stain, original magnification, 400×)

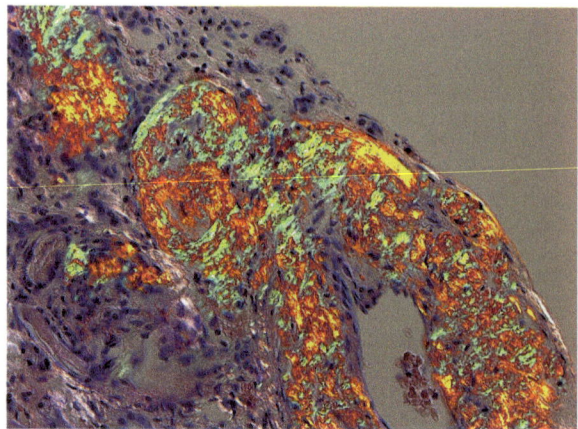

Fig. 6.60 PAS negative amyloid material within the mesangium and capillary loops (PAS stain, original magnification, 200×)

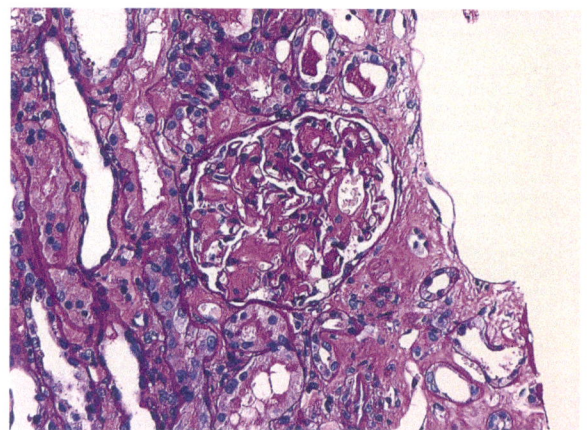

Fig. 6.61 *Pale blue* staining amyloid material within the mesangium and capillary loops (Masson trichrome stain, original magnification, 200×)

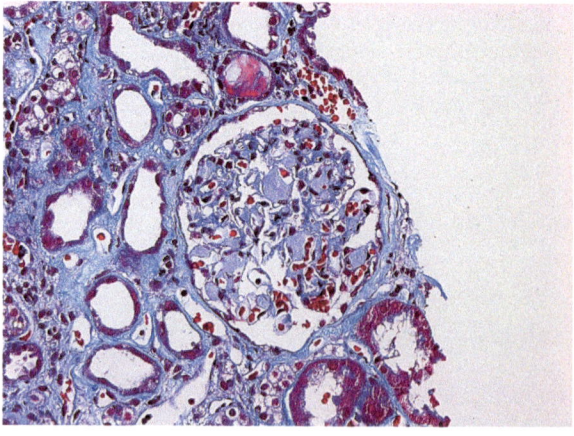

Fig. 6.62 Decreased mesangial argyrophilia within amyloid deposits (Jones Methenamine Silver stain, original magnification, 200×)

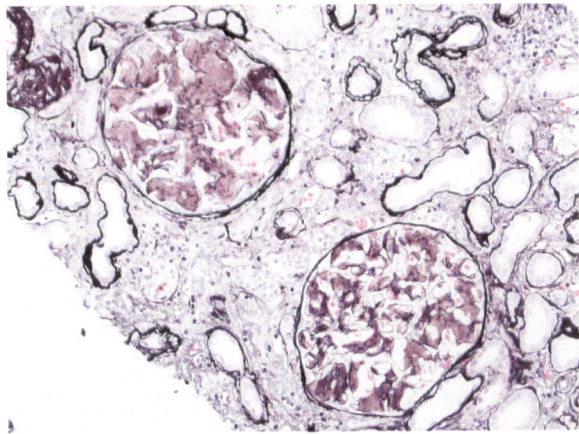

Fig. 6.63 Spicular amyloid deposition along the capillary loops (Jones Methenamine Silver stain, 400× and 600× original magnification, respectively)

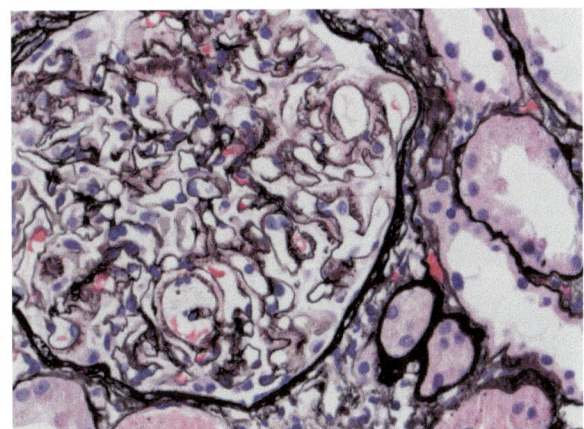

Fig. 6.64 Spicular amyloid deposition along the capillary loops (Jones Methenamine Silver stain, 400× and 600× original magnification, respectively)

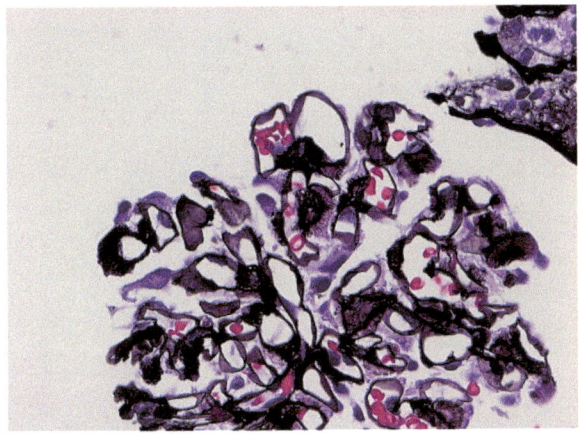

ancillary study to determine subtype. Of the three most common types of amyloid (AA, AL, LECT2), AA and AL have relatively nonspecific morphologic appearance. However, LECT2 amyloidosis has a strong association with Hispanic ethnicity and shows a predominantly tubulointerstitial pattern of deposition (Fig. 6.65) [15]. Of the rare forms, AFib amyloidosis has been reported to show extensive glomerular deposition with effacement of the underlying glomerular architecture in the near absence of vascular or tubulointerstitial deposition (Fig. 6.66) [18]. Apo A IV amyloidosis is unique in that it is reported to show large amounts of amyloidosis that are restricted to the medullary tissue (Fig. 6.67) [17]. Confirmatory immunohistochemical stains are useful in diagnosing AA and LECT2 amyloidosis and are commercially available (Figs. 6.68 and 6.69).

Immunofluorescence: IF is of paramount importance in diagnosis or exclusion of AL and AH amyloidosis. AL amyloidosis will usually show light chain

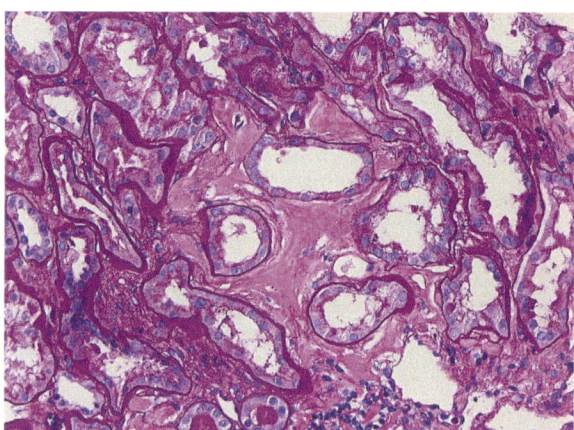

Fig. 6.65 Interstitial amyloid deposition in LECT2 amyloidosis (PAS stain, original magnification, 200×)

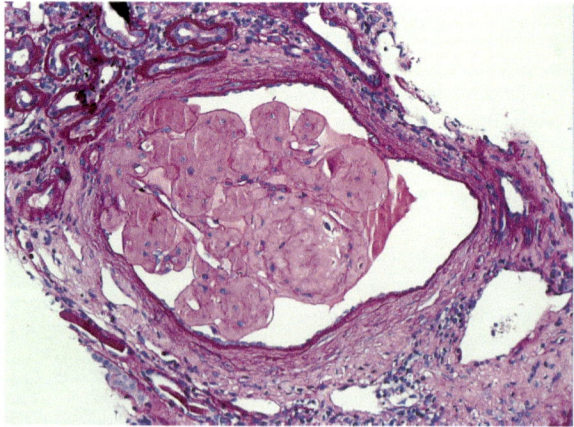

Fig. 6.66 Extensive glomerular amyloid deposition with effacement of glomerular architecture in AFib amyloidosis (PAS stain, original magnification 400×)

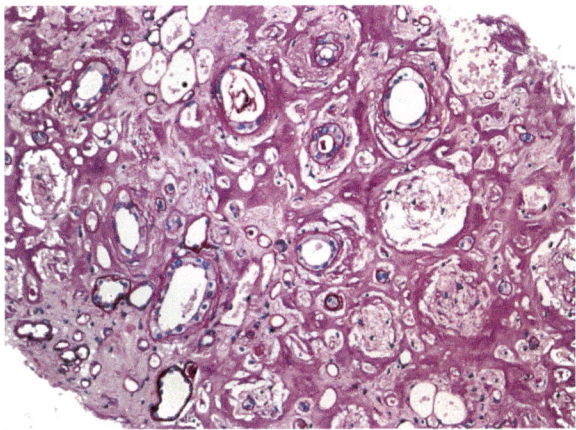

Fig. 6.67 Large quantities of interstitial, medullary amyloidosis in Apo A-IV amyloidosis (PAS stain, original magnification, 200×)

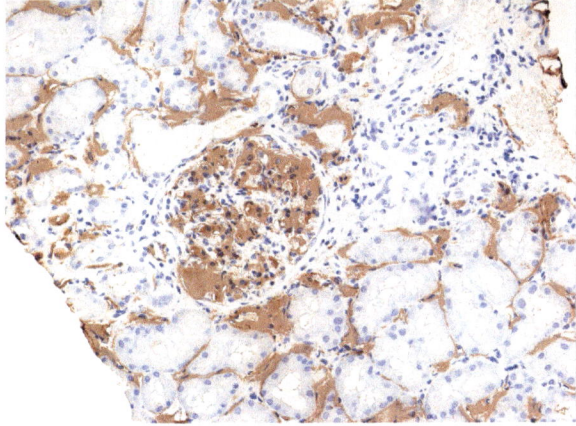

Fig. 6.68 Immunohistochemical stain for AA amyloid showing positive staining in the glomeruli, interstitium and tubular basement membranes (original magnification, 200×)

restriction by IF, most commonly lambda light chain, while AH amyloidosis shows positive IF staining for a single Ig heavy chain. The presence of additional heavy chain and complement fragment staining will occasionally be present and can produce diagnostic dilemmas. The additional presence of a single, strongly staining heavy chain raises the possibility of an AHL amyloidosis while the presence of numerous, weakly and segmentally staining immunoreactants favor nonspecific, clinically insignificant trapping. In these circumstances, confirmation of amyloid subtyping can be achieved by laser tissue capture/mass spectrometry, a methodology available in specialty labs that can be performed on paraffin-embedded tissue.

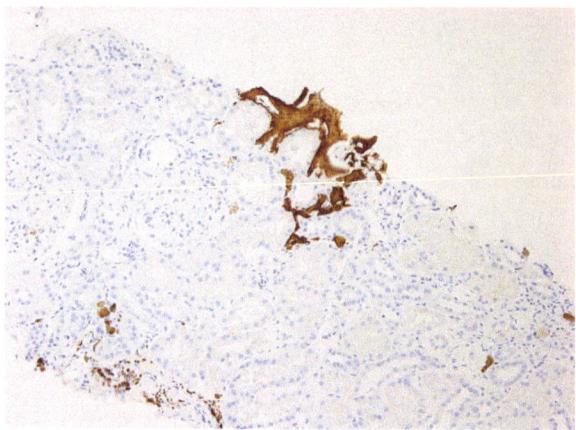

Fig. 6.69 Immunohistochemical stain for LECT2 amyloid showing positive staining within the interstitium (original magnification, 200×)

IF can also be helpful in some cases of AFib amyloidosis where amyloid deposits may show isolated staining by fibrinogen. Although this finding is useful when present, we find it to be insensitive and frequently absent in cases of AFib amyloidosis. Rarely, conflicting findings may also occur such as a positive serum A amyloid immunohistochemical stain and light chain restriction on IF. In these cases, confirmatory subtyping by laser tissue capture/mass spectrometry is of benefit.

Electron Microscopy: Generally, amyloid has a fibrillar substructure of randomly oriented, overlapping, non-branching fibrils measuring 7–12 nm (Fig. 6.70). Deposition can be seen in any portion of the renal parenchyma but is most frequently seen in the glomeruli and vasculature. In the glomerulus, deposition is often

Fig. 6.70 Electron photomicrograph showing non-branching, overlapping, randomly oriented fibril deposition in the mesangium in amyloidosis (original magnification, 25,000×)

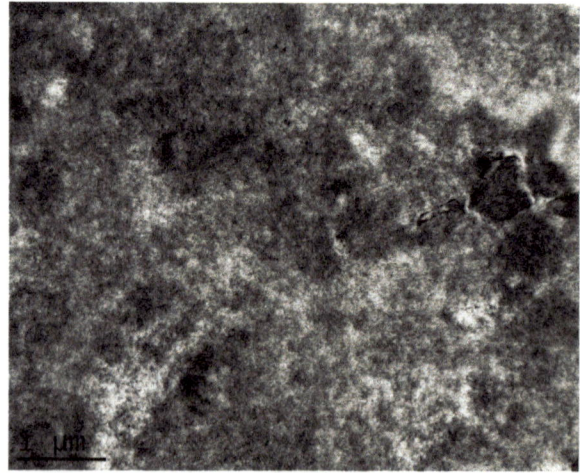

present along or within the capillary wall and in the mesangium. Rarely, extensive subepithelial deposition of amyloid fibrils will be present (this creates the spicular deposition pattern on silver stain). Only AGel amyloidosis has been reported to have specific electron microscopic findings, showing fibril organization into parallel arrays [19]. Finally, it should be noted that while electron microscopy is a highly sensitive test for the detection of amyloid, it is prone to sampling error due to the heterogeneous deposition of amyloid within the kidney and because only a very small amount of tissue is examined.

Monoclonal Gammopathy of Renal Significance

Monoclonal gammopathy of renal significance (MGRS) reflects a new approach to the classification of a group of renal diseases that are typically associated with paraproteinemia and deleterious effects on the kidney and thus increase patient morbidity and mortality. This term was first introduced in 2012 [21] and has been further expanded and defined by the International Kidney and Monoclonal Gammopathy Research Group [22]. The need for this new designation arose clinically as nephrologists and hematologists-oncologists are often confronted with a paraprotein-related kidney disease in the absence of a definable plasma cell dyscrasia or lymphoproliferative disorder. Often times, these patients only satisfy the diagnostic criteria for MGUS leaving the physician in the difficult position of deciding if treatment is merited. This new classification, MGRS, allows the pathologist to concretely define the underlying renal disease as one associated with possible increased morbidity/mortality that would benefit from treatment. The renal diseases that comprise this group include Ig-related amyloidosis (AL/AH/AHL), fibrillary glomerulopathy, immunotactoid glomerulopathy, type I cryglobulinemic glomerulonephritis, light chain proximal tubulopathy, crystal storing histiocytosis, proliferative glomerulonephritis with monoclonal IgG deposits [23], membranoproliferative glomerulonephritis with masked monotypic immunoglobulin deposits [24], monoclonal Ig deposition disease and C3 glomerulopathy with monoclonal gammopathy [25]. This is a morphologically and clinically heterogeneous group of diseases, often with complicated diagnostic criteria requiring advanced laboratory techniques and close clinicopathologic correlation with the treating physician. They can be challenging to diagnose and, as many are recently described, are in a process of evolution. Also, the decision to treat these patients with cytotoxic or other chemotherapeutic regimes is often based solely on the type and extent of renal damage seen.

Light Microscopy: Ig-related amyloidosis (AL/AH/AHL), light chain proximal tubulopathy and monoclonal Ig deposition disease have been discussed earlier in the chapter. The remaining entities included in MGRS are described in this section.

Fibrillary glomerulopathy: Shows mesangial expansion by a PAS-positive material that shows loss of argyrophilia (Figs. 6.71 and 6.72) on silver stain and

Fig. 6.71 Global mesangial expansion in fibrillary glomerulopathy (PAS stain, original magnification, 400×)

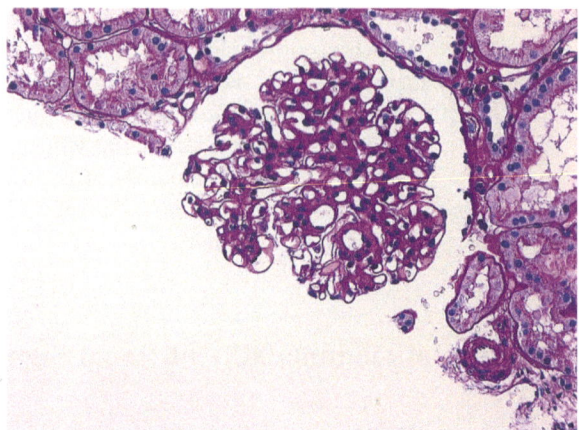

Fig. 6.72 Reduction in mesangial agyrophilia in fibrillary glomerulopathy (Jones Methenamine Silver stain, original magnification, 400×)

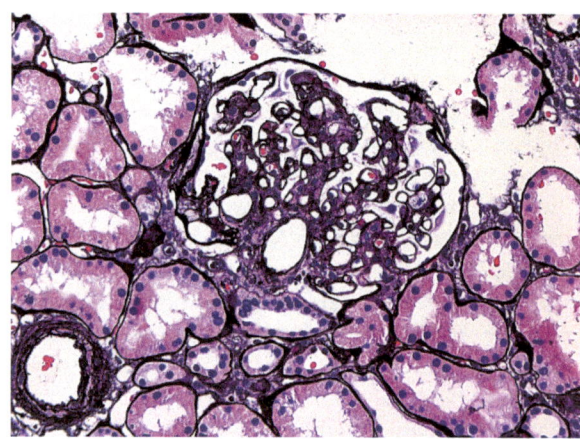

pale blue to lilac coloring on Masson trichrome stain. The deposits are Congo red negative.

Immunotactoid glomerulopathy: Can show a variety of glomerular morphologies (mesangioproliferative, membranoproliferative, or nodular glomerulosclerosis) (Fig. 6.73) as well as spike formation on silver stain and rarely crescent formation. The deposits are Congo red negative.

Type I cryoglobulinemia: Membranoproliferative glomerulonephritis with double contour formation and intracapillary hyaline thrombi (Fig. 6.74). Deposits are Congo red negative.

Crystal storing histiocytosis: Enlarged interstitial histiocytes containing eosinophilic crystalline inclusions that may also be present in podocytes and proximal tubular epithelial cells.

Proliferative glomerulonephritis with monoclonal IgG deposits: Can show a membranoproliferative, endocapillary proliferative, mesangial proliferative or

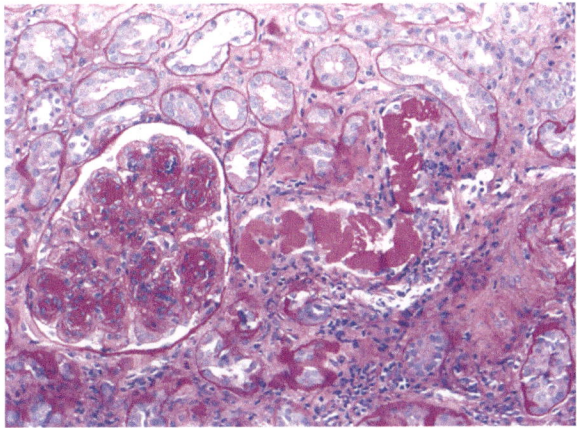

Fig. 6.73 Membranoproliferative pattern glomerulonephritis in immunotactoid glomerulopathy with adjacent intratubular light chain cast (PAS stain, original magnification 200×)

Fig. 6.74 Cryoglobulinemic glomerulonephritis with intracapillary hyaline thrombi and double contour formation (PAS stain, original magnification 400×)

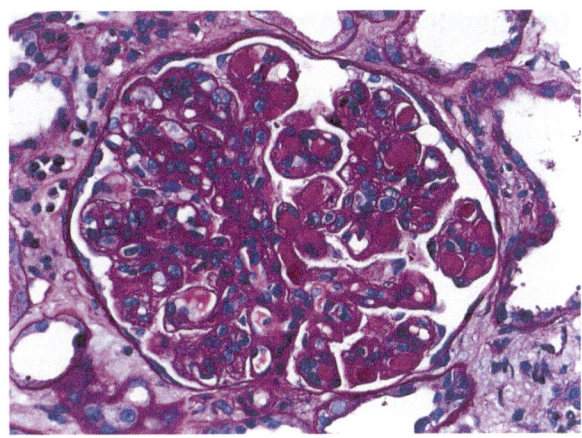

membranous morphology and may have crescent formation [23]. Congo red negative deposits.

Membranoproliferative glomerulonephritis with masked monotypic immunoglobulin deposits: Membranoproliferative pattern and may have crescent formation (Fig. 6.75) [24]. Congo red negative deposits.

C3 glomerulopathy with monoclonal gammopathy: Can show a membranoproliferative glomerulonephritis, diffuse proliferative glomerulonephritis, mesangial proliferative glomerulonephritis, necrotizing and crescentic glomerulonephritis and sclerosing glomerulonephritis [25]. Congo red negative deposits.

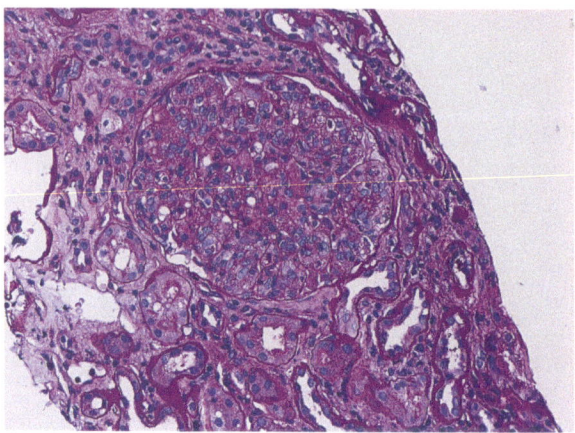

Fig. 6.75 Membranoproliferative pattern glomerulonephritis (PAS stain, original magnification, 400×)

Immunofluorescence

Fibrillary glomerulopathy: Mesangial to capillary wall, smudgy/granular staining for IgG, C3, kappa and lambda light chain (Fig. 6.76). However, it is important to note that monotypic light chain staining can occur infrequently.

Immunotactoid glomerulopathy: Capillary wall and mesangial granular staining by IgG, C3 and kappa or lambda light chain. Polyclonal light chain staining has been reported.

Type I cryoglobulinemia: Capillary wall and mesangial, granular staining for one heavy chain and one light chain, most commonly IgG3 and kappa.

Crystal storing histiocytosis: Light chain restriction of crystalline material within the cytoplasm of interstitial histiocytes and occasionally podocytes. As in proximal

Fig. 6.76 Smudgy, mesangial to capillary loop staining on immunofluorescence by IgG in fibrillary glomerulopathy (original magnification, 200×)

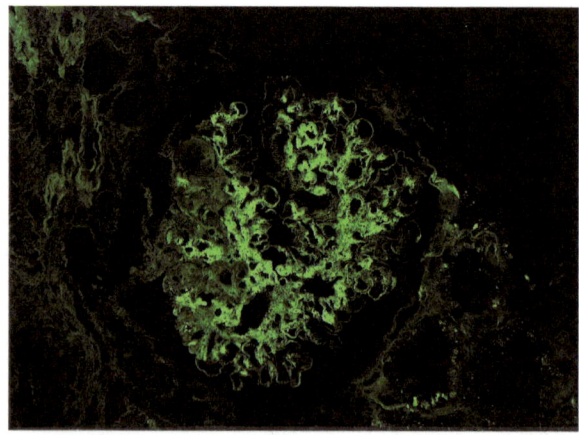

tubulopathy, routine immunofluorescence may show no staining and require paraffin-retrieved immunofluorescence including an antigen retrieval step for light chain restriction to be evident.

Proliferative glomerulonephritis with monoclonal IgG deposits: Glomerular location of deposition depends on underlying morphological pattern of injury and may be capillary wall or mesangial. Typically shows IgG3 kappa-restricted deposits but may show restriction to any IgG subclass and light chain (see Fig. 6.77).

Membranoproliferative glomerulonephritis with masked monotypic immunoglobulin deposits: Show isolated C3 deposition with negative routine immunofluorescence panel (can be easily misdiagnosed as C3 glomerulonephritis). Retrieval of paraffin-embedded tissue with an antigen retrieval step most frequently shows IgG kappa-restricted deposits but can also show IgG lambda, IgM kappa or IgM lambda-restricted deposits (Fig. 6.78). Routine staining of IgG subtypes will be negative.

C3 glomerulopathy with monoclonal gammopathy: Mesangial and/or capillary wall, granular deposits of typically IgG and either kappa or lambda. However, restriction to a single heavy chain or light chain may occur.

Fig. 6.77 **a** Granular, capillary wall deposition of kappa light chain by immunofluorescence in proliferative glomerulonephritis with monoclonal IgG deposits (original magnification, 400×). **b** Negative staining for lambda light chain by immunofluorescence in proliferative glomerulonephritis with monoclonal IgG deposits (original magnification, 400×)

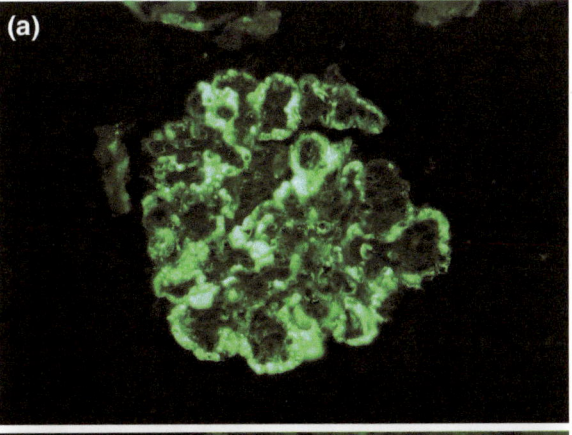

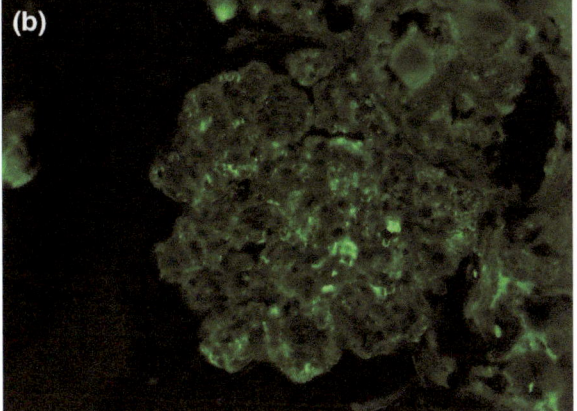

Fig. 6.78 In this case of membranoproliferative glomerulonephritis with masked monotypic immunoglobulin deposits, only scant C3 deposition was seen by routine immunofluorescence (**a**) while on paraffin-retrieved immunofluorescence (including an antigen retrieval step), intense IgG (**b**) and kappa light chain (**c**) deposits were unmasked (original magnification, 400×, 400×, and 200×, respectively)

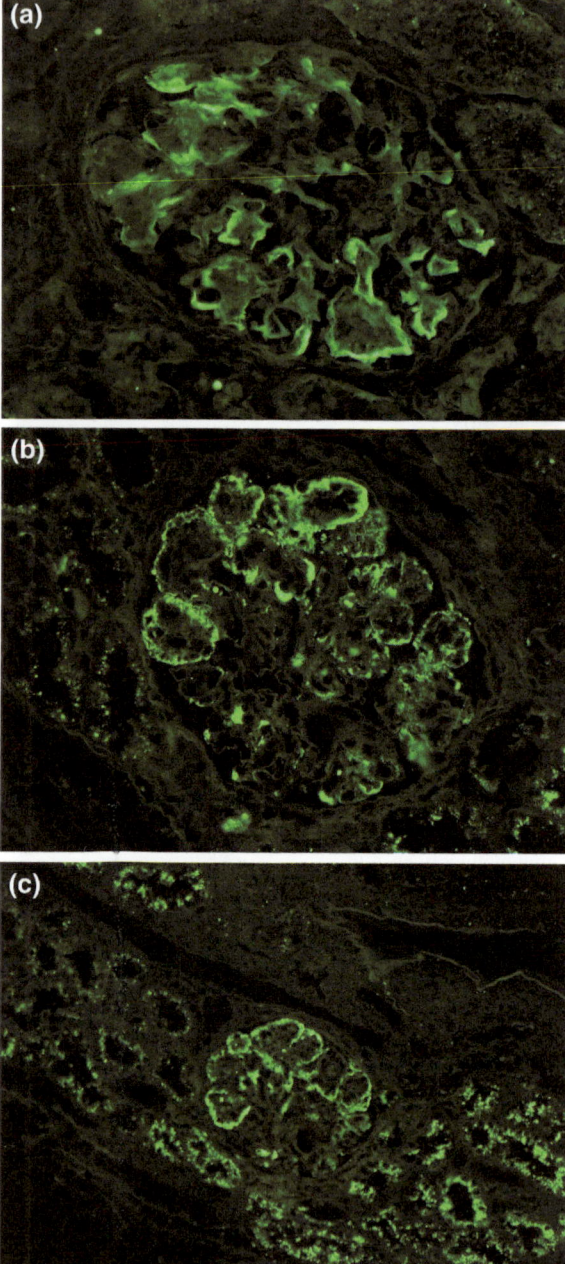

Electron Microscopy

Fibrillary glomerulopathy: Deposition of non-branching, randomly oriented fibrils without hollow-core measuring 10–30 nm within the mesangium and/or along the capillary wall (Fig. 6.79).

Immunotactoid glomerulopathy: Deposition of non-branching, hollow-core microtubules measuring >30 nm in a parallel array substructure typically within the subendothelial space and mesangium (Fig. 6.80).

Type I cryoglobulinemia: Can produce variable morphologies; however, fibrillar or microtubular deposits are most frequently seen that can involve all portions of the capillary wall, mesangium and intracapillary space. When substructure exists it tends to show straight fibrils/microtubules that are aggregated into bundles

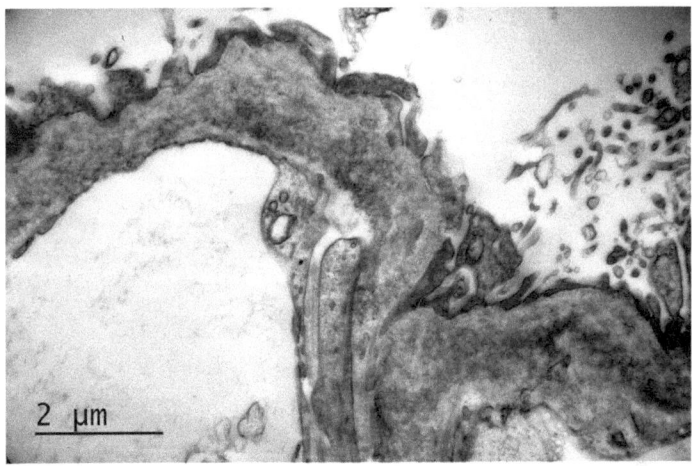

Fig. 6.79 Non-branching, overlapping, randomly oriented fibril deposition along a capillary loop in fibrillary glomerulopathy (electron photomicrograph, original magnification, 20,000×)

Fig. 6.80 Subendothelial deposits showing a microtubular substructure in immunotactoid glomerulopathy (electron photomicrograph, original magnification, 30000×)

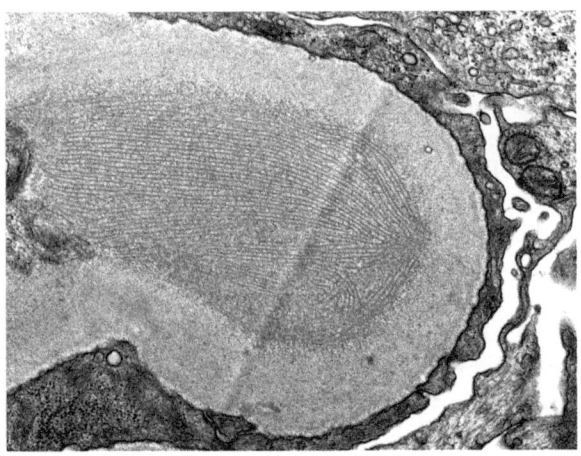

Fig. 6.81 Subendothelial (**a**) and mesangial (**b**) deposits showing a vague microtubular substructure in cryoglobulinemic glomerulonephritis (electron photomicrographs, original magnification, 20,000× and 25,000×, respectively)

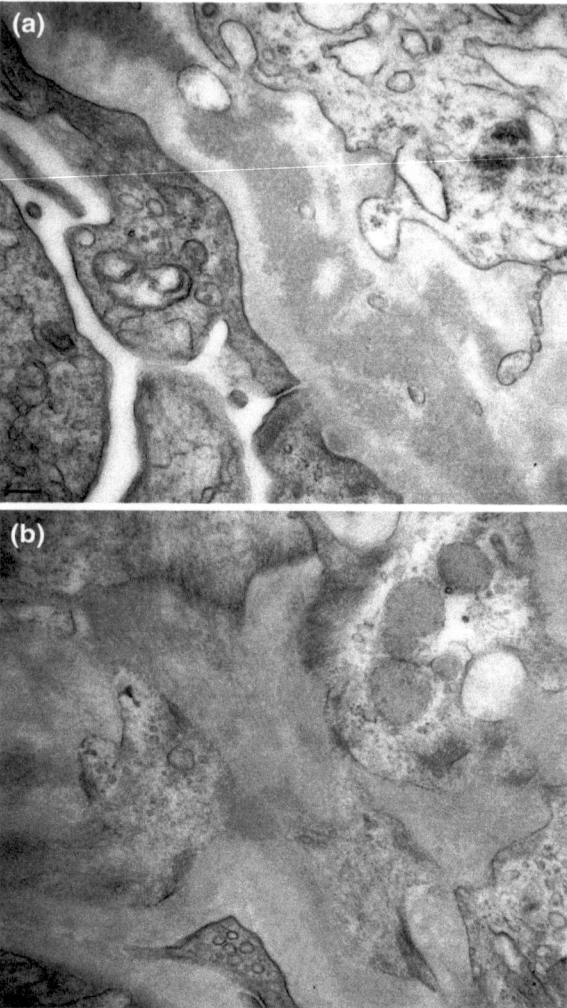

(Fig. 6.81). Capillary wall remodeling with double contour formation may be present.

Crystal storing histiocytosis: Electron-dense, crystalline structures visible within the interstitial histiocytes cytoplasm and occasionally podocyte cytoplasm.

Proliferative glomerulonephritis with monoclonal IgG deposits: Electron-dense deposits will be limited to the glomeruli; however, localization will be dependent on morphological pattern of injury. Most frequently, electron-dense granular deposits will be present in the subendothelial space and mesangium; however, subepithelial deposits are present in more than 50 % of cases. Rarely, a predominantly subepithelial pattern (membranous) of immune complex deposition can be seen. Capillary wall remodeling with double contour formation may be present.

Membranoproliferative glomerulonephritis with masked monotypic immunoglobulin deposits: Most frequently, granular, electron-dense deposits are present within the subendothelial space and mesangium and focal subepithelial deposits may be seen. Intraluminal electron-dense deposits can also be seen as well as needle-shaped, crystalline deposits. Capillary wall remodeling with double contour formation may be present.

C3 glomerulopathy with monoclonal gammopathy: Electron-dense, granular deposits within the subendothelial space and mesangium. Rarely subepithelial and intramembranous deposits may be seen. Capillary wall remodeling with double contour formation may be present.

Tubulointerstitial Nephritis, Monoclonal Light Chain Mediated

This entity morphologically mimics tubulointerstitial nephritis without glomerular changes on light microscopy. The diagnosis is made by the identification of monoclonal deposits in the tubulointerstitial compartment by IF. These deposits appear as punctate or powdery electron-dense deposits by electron microscopy and display a congruent clinical history [25].

Waldenstrom Macroglobulinemia-Associated Glomerulopathy

Waldenstrom macroglobulinemia (WM) is characterized by monoclonal IgM gammopathy secondary to lymphoplasmacytic lymphoma, a low-grade, bone marrow-based B-cell lymphoma that typically manifests some degree of plasmacytic differentiation. Patients present with clinical and laboratory features related to replacement of bone marrow by neoplastic B-cells or circulating monoclonal IgM. Renal involvement is not common. Renal biopsy on light microscopy shows glomeruli with the typical massive, intracapillary hyaline thrombi and/or large PAS-positive subendothelial deposits (Figs. 6.82 and 6.83). Cryoglobulins can be present in a subset of these patients and there is morphological overlap between these entities. The deposits are positive by IF for IgM, often for IgG and either kappa or lambda light chains (Figs. 6.84, 6.85, and 6.86). In some cases, C3 or C4 staining may also be seen. Blood vessels may show similar deposits. Electron microscopy reveals subendothelial and intracapillary amorphous electron-dense deposits (Figs. 6.87 and 6.88) [26, 27].

Fig. 6.82 PAS-positive intracapillary and subendothelial deposits (original magnification, 400×)

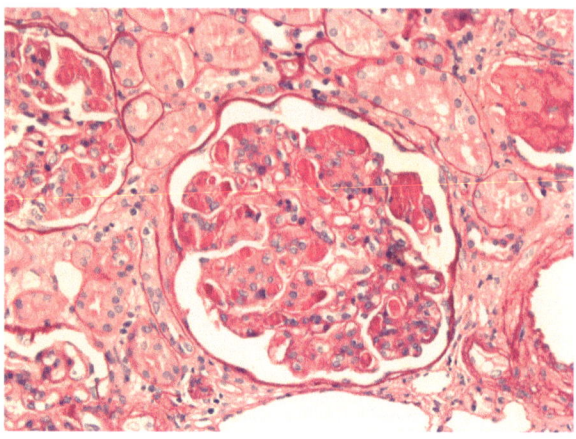

Fig. 6.83 Intracapillary and subendothelial deposits on the Jones Methanamine Silver stain (original magnification, 400×)

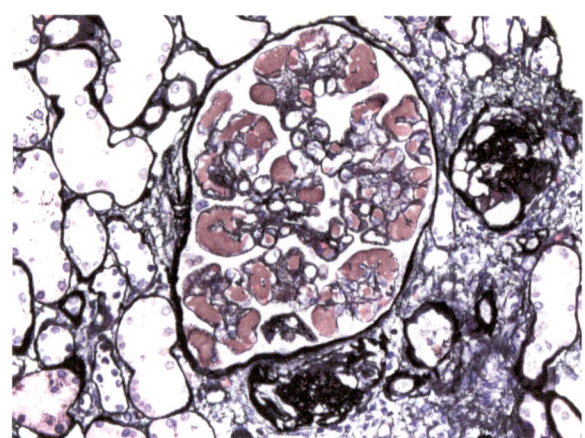

Fig. 6.84 IF showing global subendothelial and intracapillary IgM deposits (original magnification, 400×)

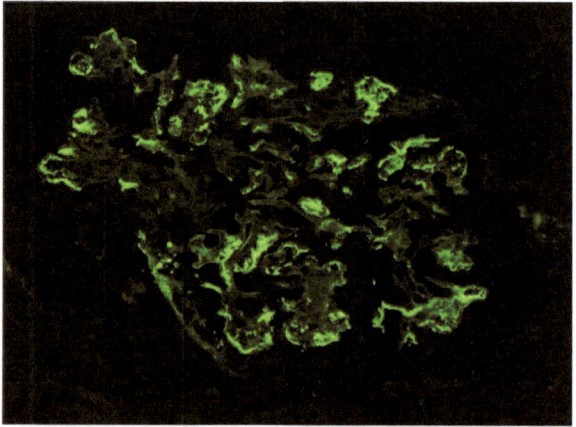

Fig. 6.85 IF showing deposits in the same distribution as IgM, which are positive for lambda light chain (original magnification, 400×)

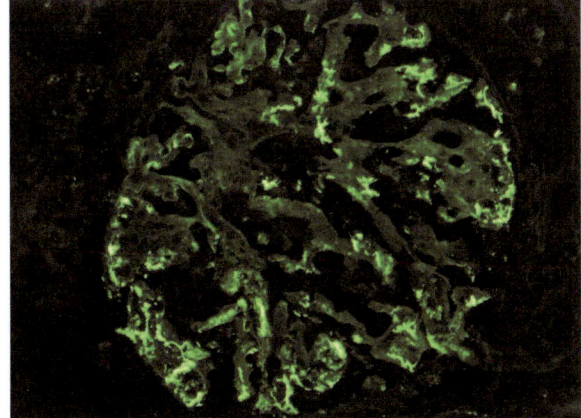

Fig. 6.86 IF showing deposits negative for kappa light chain (original magnification, 400×)

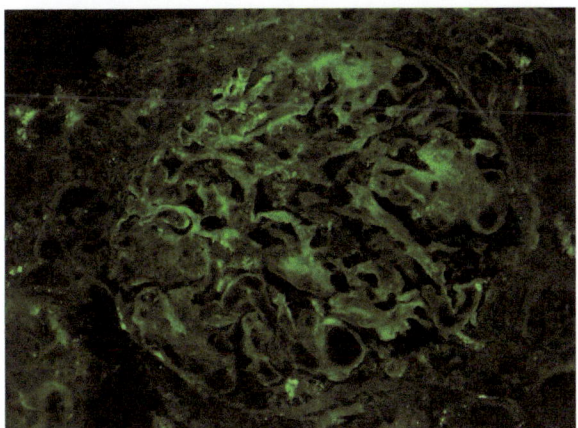

Fig. 6.87 Electron microscopy reveals subendothelial and intracapillary amorphous electron-dense deposits

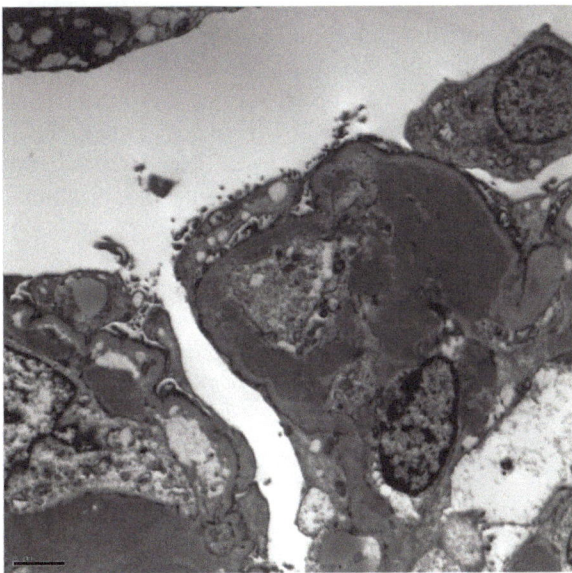

Fig. 6.88 Electron
microscopy reveals
subendothelial and
intracapillary amorphous
electron-dense deposits

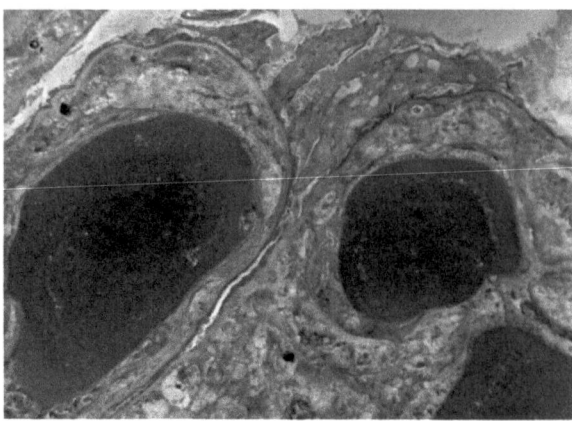

Conclusion

While reviewing a kidney biopsy from a patient with dysproteinemia, it is important
to consider the entities mentioned in this chapter, but also to include within the
differential diagnosis other potential causes of kidney injury that are not associated
with dysproteinemia. In addition, various dysproteinemia-related renal diseases can
be present in combination, e.g., one third of light chain deposition disease cases
occur together with light chain cast nephropathy. With recent advances in diagnostic testing, all the patients with dysproteinemia-related kidney disease may not
fulfill the criteria of plasma cell myeloma, and the B-cell or plasma cell clone
producing the pathogenic monoclonal protein may be relatively small. In routine
practice, a high index of suspicion for dysproteinemic-related kidney disease is
important. Ancillary staining to include at least a PAS stain in addition to standard
H&E sections is suggested when evaluating the kidney. On H&E stain it is very
difficult to appreciate the morphology of the kidney biopsy, and inclusion of a PAS
stain is often of utility in the diagnosis of subtle and early dysproteinemia-related
kidney disease. Ancillary immunofluorescence staining on pronase/proteinase
digested paraffin tissue is another useful tool when monotypic deposits are suspected but not identified by routine IF. Due to the difficulties in diagnosing various
dysproteinemia-related diseases, expert consultation by a nephropathologist may
prove valuable in aiding appropriate classification and prognostication.

References

1. Korbet SM, Schwartz MM. Multiple myeloma. J Am Soc Nephrol. 2006;17(9):2533–45.
2. Herrera GA, Picken MM. Renal involvement in plasma cell dyscrasias. In: Jennette JC,
   Olson JL, Silva FG, D'Agati VD, editors. Hepinstall's pathology of the kidney, 7th ed.
   Philadelphia: Wolters Kluwer, 2015. pp. 956–959.

3. Sanders PW, Booker BB. Pathobiology of cast nephropathy from human Bence Jones proteins. J Clin Invest. 1992;89:630.
4. Melato M, Falconieri G, Pascali E, Pezzoli A. Amyloid casts within renal tubules: a singular finding in myelomatosis. Virchows Arch A Pathol Anat Histol. 1980;387(2):133–45.
5. Herrera GA. Renal manifestations of plasma cell dyscrasias: an appraisal from the patients' bedside to the research laboratory. Ann Diagn Pathol. 2000;4(3):174–200.
6. Kapur U, Barton K, Fresco R, Leehey DJ, Picken MM. Expanding the pathologic spectrum of immunoglobulin light chain proximal tubulopathy. Arch Pathol Lab Med. 2007;131(9):1368–72.
7. Larsen CP, Bell JM, Harris AA, et al. The morphologic spectrum and clinical significance of light chain proximal tubulopathy with and without crystal formation. Mod Pathol. 2011;24(11):1462–9.
8. Herrera GA. The contributions of electron microscopy to the understanding and diagnosis of plasma cell dyscrasia-related renal lesions. Med Electron Microsc. 2001;34(1):1–18.
9. Sharma SG, Bonsib SN, Gokden N. Light chain proximal tubulopathy: expanding the pathologic spectrum with deposition of crystalline-like inclusions in glomeruli and interstitium in addition to proximal tubules. ISRN Pathol. [541075]. 2012. doi:10.5402/2012/541075.
10. Nasr SH, Galgano SJ, Markowitz GS, et al. Immunofluorescence on pronase-digested paraffin sections: a valuable salvage technique for renal biopsies. Kidney Int. 2006;70(12):2148–51.
11. Lin J, Markowitz GS, Valeri AM, et al. Renal monoclonal immunoglobulin deposition disease: the disease spectrum. J Am Soc Nephrol. 2001;12(7):1482–92.
12. Ronco PM, Alyanakian MA, Mougenot B, et al. Light chain deposition disease: a model of glomerulosclerosis defined at the molecular level. J Am Soc Nephrol. 2001;12(7):1558–65.
13. Randall RE, Williamson WC Jr, Mullinax F, et al. Manifestations of systemic light chain deposition. Am J Med. 1976;60(2):293–9.
14. Said SM, Sanjeev S, Anthony VM, et al. Renal amyloidosis: origin and clinicopathologic correlations of 474 recent cases. CJASN. 2013;8(9):1515–23.
15. Larsen CP, Walker PD, Weiss DT, et al. Prevalence and morphology of leukocyte chemotactic factor 2-associated amyloid in renal biopsies. Kidney Int. 2010;77(9):816–9.
16. Said SM, Sethi S, Valeri AM, et al. Characterization and outcomes of renal leukocyte chemotactic factor 2-associated amyloidosis. Kidney Int. 2014;86(2):370–7.
17. Sethi S, Theis JD, Shiller SM, et al. Medullary amyloidosis associated with apolipoprotein A-IV deposition. Kidney Int. 2012;81(2):201–6.
18. Gillmore JD, Lachmann HJ, Rowczenio D, et al. Diagnosis, pathogenesis, treatment, and prognosis of hereditary fibrinogen Aα-Chain amyloidosis. J Am Soc Nephrol. 2009;20(2):444–51.
19. Sethi S, Theis JD, Quint P, et al. Renal amyloidosis associated with a novel sequence variant of gelsolin. Am J Kidney Dis. 2013;61(1):161–6.
20. Sanchorawala V. Light chain (AL) amyloidosis: diagnosis and treatment. Clin J Am Soc Nephrol. 2006;1(6):1331–41.
21. Leung N, Bridoux F, Hutchinson CA, et al. Monoclonal gammopathy of renal significance: when MGUS is no longer undetermined or insignificant. Blood. 2012;120(22):4292–5.
22. Bridoux F, Leung N, Hutchinson CA, et al. Diagnosis of monoclonal gammopathy of renal significance. Kidney Int. 2015;87(4):698–711.
23. Nasr SH, Satoskar A, Markowitz GS, et al. Proliferative glomerulonephritis with monoclonal IgG deposits. J Am Soc Nephrol. 2009;20(9):2055–64.
24. Larsen CP, Messias NC, Walker PD, et al. Membranoproliferative glomerulonephritis with masked monotypic immunoglobulin deposits. Kidney Int. In press.
25. Sethi S, Rajkumar SV. Monoclonal gammopathy-associated proliferative glomerulonephritis. Mayo Clin Proc. 2013;88(11):1284–93.

26. Gu X, Herrera GA. Light-chain-mediated acute tubular interstitial nephritis: a poorly recognized pattern of renal disease in patients with plasma cell dyscrasia. Arch Pathol Lab Med. 2006;130(2):165–9.
27. Dimopoulos MA, Galani E, Matsouka C. Waldenström's macroglobulinemia. Hematol Oncol Clin North Am. 1999;13(6):1351–66.
28. Morel-Maroger L, Basch A, Danon F, et al. Pathology of the kidney in Waldenström's macroglobulinemia. Study of sixteen cases. N Engl J Med. 1970;283(3):123–9.

Index

Note: Page numbers followed by *f*, *t* and *b* refer to figures, tables and boxes, respectively

© Springer International Publishing Switzerland 2016
R.B. Lorsbach and M. Yared (eds.), *Plasma Cell Neoplasms*,
DOI 10.1007/978-3-319-42370-8